# CORES TO CLUSTERS

# ASTROPHYSICS AND SPACE SCIENCE LIBRARY

VOLUME 324

# CORES TO CLUSTERS

## Star Formation with Next Generation Telescopes


*Edited by*

M.S.N. KUMAR

*Centro de Astrofisica da*
*Universidade do Porto, Portugal*

M. TAFALLA

*Observatorio Astronómico*
*Nacional, Madrid, Spain*

*And*

P. CASELLI

*INAF, Osservatorio Astrofisico di Arcetri*
*Florence, Italy*


Springer

M.S.N. Kumar
CAUP
Rua das Estrelas, S/N
Porto 4150-762
Portugal

M. Tafalla
OAN
Alfonso XII, 3
Madrid E-28014
Spain

P. Caselli
INAF
Osservatorio Astrofisico di Arcetri
Florence 50125
Italy

*Cover images:* JHK color composite of IRAS22134+5834 ring cluster made using the UIST instrument on the 3.8m UKIRT at Hawaii (image credit: Nanda Kumar and Chris Davis) and Clerigos Tower of Porto.
Cover Design: Ricardo Samuel Reis.

Library of Congress Control Number: 2005927386

ISBN-10: 0-387-26322-5 e-ISBN: 0-387-26357-8
ISBN-13: 978-0387-26322-9

Printed on acid-free paper.



Printed in the United States of America. (SPI/MVY)

9 8 7 6 5 4 3 2 1

springeronline.com

*To the $15^{th}$ anniversary of the*
*Centro de Astrofísica da Universidade do Porto*

# Contents

# Participants

João Alves *( ESO )*
Joana Ascenso *( FCUP/ CAUP/ ESO )*
Rui Azevedo *( CAUP )*
Rafael Bachiller *( OAN, Madrid, Spain )*
Francesca Bacciotti *( I.N.A.F. - Osservatorio Astrofisico di Arcetri )*
Zoltan Balog *( Dept. of Optics & Quantum Electronics, University of Szeged )*
Henrik Beuther *( Harvard-Smithsonian Center for Astrophysics )*
Isa Brandão *( FCUP )*
Paola Caselli *( INAF-Osservatorio Astrofisico di Arcetri )*
Riccardo Cesaroni *( INAF, Osservatorio Astrofisico di Arcetri )*
Cathie Clarke *( Institute for Astronomy, London, UK )*
Carlos Eiroa *( Dpto. Fisica Teorica. Universidad Autonoma de Madrid )*
Amadeu Fernandes *( CAUP )*
Jorge Gameiro *( Centro de Astrofisica da Universidade do Porto )*
Andras Gaspar *( Univ. of Szeged, Hungary )*
Daniela Gonçalves *( FCUP )*
Yohei Harayama *( Max Planck Institute for Extraterrestrial Physics )*
Melvin Hoare *( University of Leeds )*
Susana Iglesias-Groth *( Instituto de Astrofisica de Canarias )*
Anne-Katharina Jappsen *( Astrophysikalisches Institut Potsdam )*
Ray Jayawardhana *( University of Toronto )*
Eric Keto *( Harvard-Smithsonian Center for Astrophysics )*
Nanda Kumar *( CAUP, Porto )*
Stanley Kurtz *( UNAM, Morelia, Mexico )*
Elizabeth Lada *( University of Florida )*
Charles Lada *( Smithsonian Astrophysical Observatory )*
Teresa Lago *( CAUP )*
Hugo Ledo *( Faculdade de Ciências da Universidade do Porto )*
João Lima *( CAUP )*
Michael Meyer *( Steward Observatory, The University of Arizona )*
André Moitinho *( CAAUL )*
Oscar Morata *( Department of Physics, The Ohio State University )*
Phil Myers *( Harvard-Smithsonian Center for Astrophysics )*

Antonio Pedrosa *( Centro de astrofisica da Universidade do Porto )*
Monika Petr-Gotzens *( European Southern Observatory )*
Linda Podio *( Dipartimento di Astronomia e Scienza dello Spazio )*
Rafael Rebolo *( Instituto de Astrofisica de Canarias )*
John Richer *( Cavendish Laboratory, Cambridge, UK )*
Sérgio Sousa *( CAUP )*
Mario Tafalla *( Observatorio Astronomico Nacional )*
Jonathan Tan *( ETH-Zurich )*
Paula Stella Teixeira *( Univ. de Lisboa/Harvard-Smithsonian CfA )*
Leonardo Testi *( INAF - Osservatorio Astrofisico di Arcetri )*
Silvia Vicente *( ESO / Faculdade de Ciencias da Univesidade de Lisboa )*
Sarita Vig *( Tata Institute of Fundamental Research, Mumbai, India )*
Malcolm Walmsley *( Osservatorio di Arcetri )*
Anthony Whitworth *( School of Physics & Astronomy, Cardiff University )*
Beletsky Yuri *( European Southern Observatory )*

# Organising and Sponsors

## Organising Committee

João Alves *ESO, Garching, Germany*
Paola Caselli *INAF, Arcetri, Italy*
M. S. Nanda Kumar (Convenor) *CAUP, Porto, Portugal*
M. T. V. T. Lago *CAUP, Porto, Portugal*
Mario Tafalla *OAN, Madrid, Spain*

## Local Support

Manuel Monteiro (Web page design and computer support)
Julio Carriera (computer and audio-visual support)
Nelma Silva & Elsa Martin (Reception)
Ricardo Reis & Jorge Humberto (Posters)
Filipe Pires (Audio-Visual & General Support)
Paula Fonseca (Secretary)

## Sponsoring Institutes

FCT
Fundação para a Ciência e a Tecnologia
MINISTÉRIO DA CIÊNCIA, INOVAÇÃO E DO ENSINO SUPERIOR

FUNDAÇÃO CALOUSTE GULBENKIAN

UNIVERSITAS PORTUCALENSIS

Centro de Astrofísica
da Universidade do Porto

# Foreword

It was with pleasure that CAUP became for three days the core to the cloud of star formation experts all over the world. Close to the celebration of its 15th anniversary – therefore still in the early stages of institutional evolution – we are proud of our multiple activities in Astronomy: a productive research centre, classified as "Institution of excellence" within the Portuguese research units, but also an "Institution of Public Utility" as recognised by the Government.

Fifteen years ago we choose to play a role not only in research, as expected from any research centre but also in the training of the future astronomers and the promotion of science and scientific culture. This choice is clearly stated in our by-laws and also in the multiple activities we have carried out since. Along the years we have organized on a regular basis international Workshops similar to "Cores to Clusters". Sometimes we have chosen to organize international conferences of a larger size. On other occasions the choice has been for smaller and more informal discussion meetings. Or even doctoral schools with very different objectives. In common all those meetings have always had, besides the formal registered participants, a group of informal participants, our undergraduate students of Astronomy, so eager to be in touch with the real world.

Again this time, as the Workshop sessions progressed in the quietness of the auditorium, CAUP hall was regularly enliven by its approximately 400 daily exuberant visitors – the school kids attending the sessions of the Porto Planetarium, also a responsibility of CAUP. This is the result of a unique choice – to be scientifically responsible for the town's public planetarium. In exchange for a shared beautiful building in which conception and design we were involved right from the beginning. This common installation was in fact our proposal. In our view this structure fully optimizes complementary synergies – the planetarium benefiting from the proximity of a research institute in Astronomy that guarantee for the quality, scientific level and actuality of the contents of the public sessions. The research institute benefiting from a special stage for its activities related with public awareness of science. The practice of almost 15 years has proven the choice to be a very positive one and we are very proud of it.

CAUP was for a few days the centre of live discussions on star formation theories and observations as a result of the effort of various people with special emphasis for the organizers Joao Alves, Mario Tafalla, Nanda Kumar and Paola Caselli. Naturally Nanda Kumar being the local organizer had the big share of work. CAUP staff also made a splendid contribution for the smooth run of the meeting. And last but not the least the sponsors of this meeting deserve acknowledgment since they believed in our proposal of organizing a non-expensive productive meeting. Fundação para a Ciência e Tecnologia, Fundação Calouste Gulbenkian (Serviço de Ciência) and Reitoria da Universidade do Porto provided the needed support for the keynote speakers, coffee breaks and in part this book.

This was also a very timely meeting on a topic of high interest for CAUP researchers in both Stellar Astrophysics and the Cosmology, our two broad research areas. Especially when several major telescope facilities in the infrared, milimeter and radio bands are being built and will become fully operational within a decade. These missions are expected to play a major role towards our understanding of the star and planet formation, which, indeed is addressed, as one of their primary scientific goals. "Cores to Clusters" workshop aimed at discussing the quests of star formation to be investigated with several of new generation telescopes such as Spitzer, ALMA, VLTI, SMA, CARMA and HERSCHEL. This book includes review type articles on star formation written by leading researchers and other more technical papers on related current issues. The authors of the articles and the editors have made their best effort to bring out this book. I am quite confident it will be a useful book to all young researchers in the subject of star formation.

Teresa Lago
Director
Centro de Astrofísica da Universidade do Porto
Porto, 20th April 2005

Workshop Photo

I

# REVIEW ARTICLES

# PHYSICAL PROPERTIES OF PRESTELLAR CORES

Malcolm Walmsley
*INAF– Osservatorio Astrofisico di Arcetri*
walmsley@arcetri.astro.it

**Abstract** Prestellar cores are thought to be the ultimate phase in the development of a gravitationally bound condensation in a molecular cloud prior to the formation of a star. In this review, I discuss the basic processes governing the physical characteristics of such cores as well as the techniques used to determine such parameters as density, temperature, and magnetic field strength.

## Introduction

Our realisation that stars are presently forming in the Milky Way is relatively recent. And the realisation that stars form in molecular clouds is more recent still. Nevertheless, it is probably worth pointing out that our belief that the latter statement is correct is based to a considerable extent on the finding in the late 1960's and early 1970's (Becklin & Neugebauer 1967, Kleinmann & Low, 1967, Harper & Low, 1971) that embedded in molecular cloud material close to the Trapezium cluster was another cluster of powerful infrared "protostars". Slightly later, the molecular cloud emission was mapped (e.g Harvey et al. 1974) and it was also realised that the molecular cloud in this vicinity consists of several hundred solar masses of gas in an area roughly 0.5 parsec across. In other words as we are now aware, the molecular cloud mass is comparable to that in the Trapezium cluster and is spread over similar dimensions. Thus, understanding star formation is intimately linked to understanding molecular cloud physics.

Fortunately perhaps, later developments occurred on somewhat longer time scales. But one thing that happened was that we discovered that the dense clump in Orion as well as many other "cores" were part of much larger structures which got named giant molecular clouds or GMCs with masses anywhere from $10^4$ to $10^7$ solar masses. Apparently, the first act on the way to forming stars was to form a GMC (see e.g. Solomon et al. 1979). These GMCs were much less dense (a few hundred $H_2$ cm$^{-3}$ as compared to a few times $10^5$) than the region surrounding the Orion infrared sources but extended over typically

50 parsec and seemed to be gravitationally bound. A few percent of the mass however was in the form of high density (above $10^4$ cm$^{-3}$) cores or clumps (as in the Orion case) and these became a focus of attention. As time went on, it became clear that young pre-main-sequence stars (T Tau's) also probably had their origin in cores and sometimes formed in a more isolated mode than the IR sources close to Orion.

All of this caused people to rethink their ideas about star formation in general and star formation rates in particular. In particular also, people started in the 1970's to think hard about molecular clouds and their evolution. This brief review cannot cover all of these studies but it is useful to point out some of the many good reviews that are dispersed in the literature. In particular, I note the excellent overview by Evans (1999) , the "classical" review of the theory by Shu et al. (1987) and the articles by Myers, by McKee, and by Blitz & Williams in Lada & Kylafis (1998). A detailed discussion of the topics discussed here is also given in the book by Stahler & Palla (2004).

In this article, I will only consider some "basic concepts" concerning star formation (in section 2) as well as giving a brief summary of the methods one uses to determine physical parameters in the clouds or "cores" where star formation takes place (section 3).

## Basic Ideas

Much of our thinking about star formation is based upon the fact that a cloud of density $\rho$ and temperature T will be gravitationally unstable against collapse if the mass exceeds a critical value $M_J$ (the Jeans mass) and if thermal pressure is the only form of support (no magnetic field or turbulence). The Jeans mass is (see e.g. Evans, 1999):

$$M_J = (\frac{\pi k T}{\mu m_H G})^{1.5} \rho^{-0.5} = 18 T^{1.5} n^{-0.5} M_{\odot} \qquad (1)$$

where $n$ is the molecular hydrogen density (in cm$^{-3}$), $\mu$ is the mean mass per $H_2$ in amu, and $G$ is the gravitational constant.

This result is quite fundamental for a number of reasons but in particular because it implies that for temperatures and densities typical of the molecular cloud cores associated with star formation (say $T$=10 K and $n = 10^4$ cm$^{-3}$), the Jeans mass is around $6M_{\odot}$. To within a factor of a few, these regions have a "reason" to produce objects of a stellar mass. Indeed, there is a "knee" in the initial mass function at around $0.5M_{\odot}$and this may be related to the Jeans mass in the medium where most stars form.

Moreover, we have reason to suppose that there exist cores where gravitation and thermal pressure are in balance to the extent that some may be in hydrostatic equilibrium. One very pretty example is the "Bok globule" Barnard 68 studied by Alves et al., 2001 where they found that the density distribution (more precisely the column density distribution) corresponded well to that expected for an isothermal equilibrium sphere. In fact, this object appeared to be consistent with the theory worked out by Ebert, 1957 and Bonnor, 1956 in the 1950's. The technique used here depends upon the assumption that the grains in these high density cores extinct background starlight according to the same laws obeyed by the more well studied grains in the diffuse interstellar medium. Given that the grains in question are thought to be covered with ice mantles of frozen water and CO, it is not obvious what one should expect and it is comforting that both theoretical (Ossenkopf & Henning 1994) and observational work (Bianchi et al. 2003) suggest the assumption is reasonable.

There is not only a mass associated with a given $T-\rho$ combination. There is also a timescale which is the time on which collapse would take place without pressure support. This is the free–fall time $\tau_{ff}$ which is for a cloud of density $\rho$ , the minimum timescale on which dynamical changes take place. We have:

$$\tau_{ff} = (\frac{3\pi}{32\, G\, \rho})^{0.5} = 3.4\; 10^5\, n_4^{-0.5} yr \tag{2}$$

where $n_4$ is the molecular hydrogen density in units of $10^4$ cm$^{-3}$. Note that this only depends upon the density and that higher density regions evolve faster. Thus intuitively, one expects that in star formation, the first step (at lowest density) will be the slowest and will determine the timescale of, say, the protostellar collapse. Thus once a core is formed, one might expect star formation to last a few hundred thousand years. This is in fact roughly what one finds.

Another thing to notice is that dividing the Jeans Mass by the free-fall time gives one (dimensionally at least) a mass accretion rate. One also sees doing this that the "typical" accretion rate derived in this way *only depends on the temperature*. On the other hand, the rate at which a protostar evolves is very tightly linked to the rate at which it accretes mass and so we come to the rather surprising conclusion that protostellar evolution depends critically on the temperature of the clump or core from which a prostar forms. This obviously is a very loose argument but a more sophisticated discussion (see Shu et al. 1987) shows that in many circumstances, it is a useful approximation to the truth. Clearly however, it only holds in situations where thermal pressure is the main force counterbalancing gravitation and equally clearly, this is often not the case.

The two other main candidates for preventing gravitational collapse are the magnetic field on the one hand and turbulence on the other. I will consider them briefly here but note that the review by MacLow & Klessen (2004) gives an excellent overview of the topic. Concerning turbulence, one should stress that it has been known from the first observations of molecular clouds that they are permeated by turbulence and that the turbulence is normally supersonic. This is shown by observations of line broadening in molecular clouds where one finds that, in most cases, the observed linewidth cannot be explained by thermal broadening. In fact the observed line widths tend to correlate with size with relatively narrow lines being found towards small dense cores with densities of $10^4$ cm$^{-3}$ or more. An example is the B68 core mentioned earlier which appears to be in hydrostatic equilibrium and where turbulence makes a relatively modest contribution to the broadening. Thus thermal (as opposed to turbulent) pressure seems in this case to dominate.

In other cases however and, in paticular in regions like Orion where high mass stars are forming, the turbulent pressure may play a fundamental role opposing gravity. If turbulent support is playing a role in preventing gravitational collapse, it's behavior may be similar to that of thermal pressure. Thus, one can have an effective Jeans mass where instead of the sound (or thermal) velocity, one substitutes the RMS turbulent velocity. In this view of things, turbulence just tends to act like a higher "temperature" thus causing a larger Jeans mass and accretion rate. Thus one might naively expect higher mass stars to form in "turbulent" cores. The observational evidence on this point is mixed however and it may be interesting in the future to compare observations of the stellar initial mass function (IMF) in molecular clouds with different levels of turbulence.

The other contender for preventing collapse is the magnetic field and here, the main difficulty is knowing how large the field is in the dense cores from which star formation takes off. I discuss this below but note here that a critical characteristic of the field is that even in molecular cloud cores where the degree of ionization $n_e/n_h$ (where $n_e$ is the electron density and $n_h$ the hydrogen density) approaches $10^{-8}$, one expects the field lines to be attatched to the gas. In other words, the field is "frozen in" and the equations of magnetic hydrodynamics (MHD) hold. Then also, one expects the magnetic energy in a contracting cloud to have the same dependence on density as the gravitational energy ($B^2 R^3$ varies as $1/R$ if flux freezing holds). Thus if initially, magnetic pressure is unable to prevent collapse, this will also be true as the core contracts. More formally, one can show that a magnetic flux $\Phi$ can support a cloud against contraction if the cloud mass does not exceed a critical value $M_{cr}$ given by (see Shu et al. 1987, Stahler & Palla, 2004) :

$$M_{cr} = 0.13 \frac{\Phi}{G^{0.5}} \simeq 2.5 \frac{B}{30\,\mu G} (\frac{R}{0.1pc})^2 M_\odot \tag{3}$$

One should note that this is in essence a condition on the cloud column density $M/R^2$ relative to the magnetic field. Another way of putting this is to say that for any observed cloud column density, there is a critical magnetic field and in fact, the estimates we do have (e.g. Crutcher, 1999) in fact suggest that fields are close to that critical value.

Conversely, in a model where magnetic field does prevent collapse initially, it will continue to do so until nature finds some way of separating field lines from the gas to which they are attatched. The most carefully studied mechanism of this latter type is the so-called "ambipolar diffusion" where gravitation slowly allows neutral hydrogen molecules to drift relative to both the field lines and the charged component of the core. The review by Shu et al. 1987 (but also Stahler & Palla, 2004) provides a good description of this. It is worth noting that ambipolar diffusion can be modulated either by the positively charged ions (such as $HCO^+$) or by the negatively charged grains (it turns out that most grains are negatively charged in prestellar cores).

For the case where the ions dominate, the timescale for ambipolar diffusion is proportional to the product of a drag coefficient $\gamma_{in}$ for momentum transfer in collisions between ions and neutrals and the ratio of ion and neutral densities $\rho_i/\rho_n$. It thus depends sensitively on the degree of ionization in the largely neutral plasma. This in turn is determined by cosmic ray ionization of hydrogen and helium on the one hand and recombination of ions either with electrons or with negatively charged grains (e.g. McKee, 1989). Hence, rather surprisingly, one has the situation where in principle at least, the rate at which stars form can depend on the degree of penetration of galactic cosmic rays into the molecular cloud. However, as far as we can tell, cosmic rays do penetrate as evidenced by the existence of molecular ions such as $HCO^+$ and $N_2H^+$ in the high density cores of clouds. Assuming therefore a "normal" cosmic ray flux similar to that measured at earth, one finds that the ambipolar diffusion time is typically an order of magnitude larger than $\tau_{ff}$ under conditions typical of observed cores. Thus, if indeed magnetic fields are larger than the critical value defined by equation 3 , star formation is a slow process (see Palla & Stahler 2002 for a discussion of observed star formation rates).

## Determining Physical Parameters of Star Forming Regions

The previous discussion has hopefully demonstrated that it is important to determine accurately the physical parameters of dense cores in order to judge their degree of evolution and to decide whether they in fact will eventually collapse to form a star. These parameters include mass and size as well as the density and temperature distribution, magnetic field, and ionization degree. In this section, we give a brief description of current approaches to determining these parameters as well as some comments on their degree of reliability. In some sense, this amounts to a list of *caveats* or warnings and so I should stress that despite appearances, it seems likely that our current estimates are *qualitatively* correct and in some cases, this is all one need to know. We start with a brief discussion of mass determination of cores.

## Mass determinations

Mass determinations for cores are complicated by a number of factors. One is that much of the mass of a core is in it's exterior (a spherically symmetric density distribution going as $1/r^2$ has mass proportional to outer radius) and that, as a consequence, the definition of an edge to a core becomes of importance. However, this is a rule not clear and also, spherical symmetry is seldom if ever observed. This latter point is fundamental and means that *results obtained for cores using spherically symmetric codes (no matter how sophisticated) should be treated with suspicion.* And last but by no means least, all mass determinations have to contend with the fact that the main constituent of cores which is molecular hydrogen is effectively "invisible". One therefore uses substitute tracers such as the dust whose abundance one believes to be proportional to the gas density. This is all one can do but it is not an ideal situation!

An added complication when considering molecular line measurements is that there is abundant evidence that most of the heavy element content of the gas has been frozen out onto dust grain surfaces in the high density core nucleus ($n_H$ above $10^5$ cm$^{-3}$, see Tafalla et al. 2004, Bacmann et al. 2002). The consequence of this is that people have used either the mm/submm emission (e.g. Young et al. 2003) or the NIR/MIR absorption (Alves et al., 2001,Bacmann et al. 2000) to determine the mass distribution within cores. The mm dust emission has been the most widely used technique and the results show rather clearly the departures from spherical symmetry mentioned earlier. The mm emission has the drawback that the dust emissivity at these wavelengths has a factor of 2 uncertainty (see eg Bianchi et al. 2003, Kramer et al. 2003) and may be different in the cold interior (where the material is essentially all frozen out) from in the somewhat less frigid exterior where water is frozen out but CO is in the gas phase. Another complication (see Evans et al. 2001, Galli et al. 2002) is that the temperature gradient in the core means that the submm maps are biased towards the exterior.

Mass determination using NIR absorption (which avoids some of the above problems) will be discussed elsewhere in these proceedings. We note here that this technique like its MIR counterpart (Bacmann et al. 2000) has the advantage of measuring the IR extinction which should be proportional to the dust column density. It is clear that results from the Spitzer Space Telescope are going to allow considerable progress in this type of study and may finally give results which one can use with some confidence.

## Temperatures

Concerning the temperature, it is worth first stressing that one usually discusses either the gas temperature or the dust temperature and the two are not, as a rule, identical (see Galli et al. 2002) although they are usually close to one another at densities above $10^5$ cm$^{-3}$. In general, the dust temperature distribution is fundamental for understanding the energetics within a cloud. For example, a temperature peak towards cloud center suggests the presence of an embedded protostar whereas temperature decrease towards cloud center is expected in the case of heating solely by the external interstellar radiation field.

The dust temperature distribution is usually inferred from the spectrum of radiation emitted in the mm to far infrared wavelength region (see Ward–Thompson et al. 2002) which can typically be fitted with a modified black body. In practise, temperature gradients are present and it is often better to fit to a model of the temperature distribution. Then in the high density nucleus, as mentioned above, one expects gas and dust temperatures to approach one another and one can use the gas temperature as a surrogate for the dust.

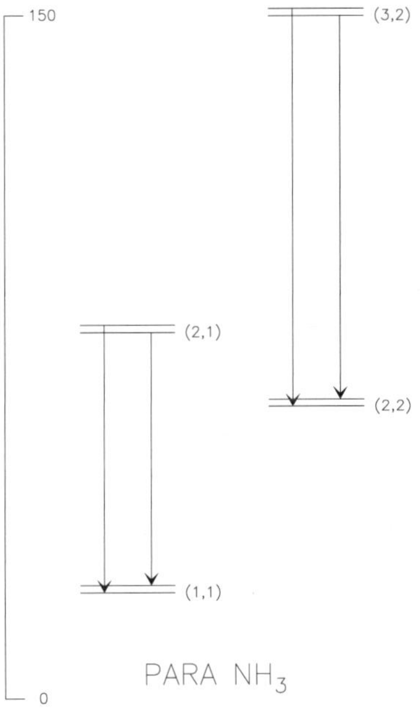


*Figure 1.* Energy level diagram for the lowest levels of the para form of $NH_3$. The rotational (J,K) levels are split due to inversion by an energy corresponding roughly to 1 K (not to scale on the figure). The bar on the left corresponds to 150 K. The arrows show allowed rotational transitions. One notes that transitions between the K=2 and K=1 ladders are forbidden.

The gas temperature has usually been inferred from molecular line studies. In particular, one observes molecules such as symmetric tops where the radiative selection rules are such that line intensity ratios are close to those expected under local thermodynamic equilibrium (LTE). The classical example is ammonia ($NH_3$) where radiative transitions between differing K-ladders are forbidden (see Ho & Townes, 1983, Stahler & Palla, 2004). Fig. 1 shows the energy level diagram for the lowest levels of the para (H spins anti–parallel) form of ammonia. One sees that the inversion splitting has the consequence that there are long wavelength ( 1.3 cm) transitions which allow us to probe the column density in different rotational levels using the same telecope and experimental conditions (hence relative calibration is reliable). The fact that there are no allowed radiative transitions between ladders has the consequence that in many situations, the population ratios of levels of differing K are approximately given by the Boltzmann law. Thus, the population ratio of the (1,1) and (2,2) metastable inversion levels is in this case :

$$\frac{n(2,2)}{n(1,1)} = \frac{g_2}{g_1} \exp\left(-\Delta E/(kT)\right) \tag{4}$$

where $\Delta E$ is the energy difference between the levels and the statistical weights of (2,2) and (1,1) are $g_2 = 5$ and $g_1 = 3$ respectively. Numerically in this case $\Delta E/k$ is 41 K and thus for temperatures of order 10 K, the (2,2) level is much less populated than (1;1) and consequently, the (2,2) line is much less intense than (1,1). In reality, the above is only an approximation because the excited non–metastable levels (e.g. (2,1) and (3,2) on Fig. 1) have, for typical densities, populations much smaller than in LTE. This causes corrections to equation 4 which are discussed by Danby et al. 1988.

Another measure of temperature is the intensity in optically thick thermalised transitions such as those of $^{12}$CO(1-0). Here, one merely applies the black body formula and assumes thereby that the telescope "beam" is filled. The combination of low dipole moment and high abundance makes CO ideal for this purpose (the level populations are determined by collisions and hence LTE holds). A drawback here is that one is essentially measuring the temperature of an outer surface of the cloud where the CO(1-0) optical depth is of order unity. Thus, if the cloud is heated from the outside, there is always some doubt about how representative the values derived are.

## Densities

Mass or column density determinations can always be converted to density determinations with some assumption about source geometry. However, the fact that most observed dense cores are not circularly symmetric renders the assumption of spherical symmetry inadequate. It nevertheless is often used *faute de mieux* and there may be cases (e.g. B68 of **?**) where it is valid. Even then, the *caveats* discussed above concerning mass determination render density estimates of this type questionable.

The alternative is to profit from the fact that LTE is rarely reliable using density sensitive line ratios in a manner analogous to that which has been traditionally used to estimate the electron density in HII regions. This can be done for example observing the differing rotational transitions of linear (or quasi–

linear) species such as CS, HCN, or $H_2CO$. Here, one exploits the fact that radiative transition rates are a strong function of frequency and hence one can use the relative intensity of higher frequency lines as a measure of the collision rate. This has the great advantage that one is directly measuring the $H_2$ density as opposed to some surrogate such as the dust. The disadvantages are that one has to correct for trapping (high optical depth has roughly the same effect as high density) and that real clouds are highly inhomogeneous and so one is forced to consider cloud structure. Evans (1999) gives a useful summary of the techniques and density probes commonly used (but see also Keto et al. 2004).

## Magnetic Field and Ionization Degree

As noted earlier, the magnetic field is a critical parameter for understanding star formation. It is also extremely hard to measure. The most attractive approach is to use the Zeeman effect but this has proved to be extremely hard in particular for cold dense regions. One needs in practise to have a probe with an unpaired electron such as OH or CN and one is hampered by the fact that the splitting does not increase with frequency. Thus, Zeeman effect is most easily observed at *low frequency* and indeed large low frequency arrays such as SKA show promise for the future. Presently, some of the best results have been obtained with the Arecibo telescope observing OH (see e.g. Crutcher & Troland 2000) with a beam of roughly 3 arc minutes or roughly 0.1 parsec at the distance of Taurus (too coarse for the dense core nuclei).

On the other hand, from the polarization of the dust continuum emission, it is clear that (see Crutcher et al. 2004) magnetic fields are capable of aligning dust particles in pre–stellar cores. There have been attempts to infer the magnetic field strength from the polarization angle dispersion in cores applying the Chandrashekar–Fermi formalism whereby the measured dispersion is ascribed to turbulence (the strength of the turbulence being estimated from the broadening of molecular transitions). It is unclear however how applicable these techniques are in the presence of a structured magnetic field (Gončalves et al. 2004). The observations show a tendency for lower polarization towards the dust intensity peaks and it is unclear whether this is due to increased turbulence or a tangled magnetic field.

In the case that cores are sub–critical (mass less than the critical value of 3), it becomes important to know the ambipolar diffusion timescale and hence the ionization degree. Determining this parameter has been a considerable headache for a number of reasons (see Caselli et al. 1998, Williams et al. 1998). The main technique used here has been to measure abundances for the observable molecular ions and use these to infer the electron density. Thus for example, under some circumstances, one expects that $HCO^+$ becomes a major ion and in such cases, assuming charge neutrality, one has a useful lower limit to the electron abundance (the grain contribution to the negative charge is negligible unless grains are very small). The catch here is that chemical models suggest that atomic ions (including $H^+$ and $S^+$) which one cannot directly observe may be as abundant or more abundant than $HCO^+$. Moreover, in the dense regions where CO is depleted, $HCO^+$ also disappears from the gas phase leaving either $H_3^+$(or one of it's deuterated forms) or $H^+$ as major ions (see Walmsley et al. 2004). Neither $H^+$ or $H_3^+$ can presently be observed from high density prestellar cores and the best hope may be the direct observation of $H_2D^+$( Caselli et al. 2003) and $D_2H^+$( Vastel et al. 2004). However, in all of these cases, one is forced to use chemical models to account for the unobserved ions and the uncertainties in such models are large (see Caselli's contribution).

## Future Prospects

Where can one expect future progress in this field? Although predictions are very hard to make, it is clear that the instrumentation which one can expect in the next decade will allow a wealth of new data to be obtained. In particular, ALMA, SKA, and new cameras in the mid and near IR will permit sensitivity which we can presently only dream about. We will see, for example, details within the high density nuclei of prestellar cores which were previously inaccessible. The most important conceptual step forward however would be to get quantitative data on the magnetic field distribution and whether this will happen quickly is debatable.

## Acknowledgments

Thanks are due to Paola Caselli and Daniele Galli for their comments on an earlier version of this manuscript. I would also like to thank Nanda Kumar and the Porto Observatory for their hospitality.

## References

Alves, J. F. Lada, C. J., & Lada, E. A. (2001). *Nature*, 409:159–161.

Bacmann, A., André, P., Puget, J.-L., Abergel, A., Bontemps, S. & Ward-Thompson, D. (2000) *A&A*, 361:555

Bacmann, A., Lefloch, B., Ceccarelli, C., Castets, A., Steinacker, A., & Loinard, L. (2002) *A&A*, 389:L6

Becklin E. E., & Neugebauer G. (1968) *ApJ*, 147:799

Bianchi, S., Goncalves, J., Albrecht, M., Caselli, P., Chini, R., Galli, D. & Walmsley, M. (2003) *A&A* 399:L43

Bonnor, W. B. (1956) *MNRAS* 116:351

Caselli, P., Walmsley, C. M., Terzieva, R. & Herbst, E. (1998) *ApJ* 499:234

Caselli, P., van der Tak, F. F. S., Ceccarelli, C. & Bacmann, A. (2003) *A&A* 403:L37

Crutcher, R. M. (1999) *ApJ* 520:706

Crutcher, R. M., Troland, T. H.(2000) *ApJ* 537:L139.

Crutcher, R. M., Nutter, D. J., Ward–Thompson, D. & Kirk, J. M. (2004) *ApJ* 600:279

Ebert, R. (1957). *Zeitsch. für Astrophys* 42:263

Danby, G., Flower, D. R., Valiron, P., Schilke, P. & Walmsley, C. M. (1988) *MNRAS* 235:229

Evans, N. J. II (1999) *ARA&A* 37:311

Evans, N. J. II, Rawlings, J. M. C., Shirley, Y. L & Mundy, G. (2001) *ApJ* 557:193

Galli, D., Walmsley, M., & Gonc̆alves, J. (2002) *A&A* 394:275

Gonc̆alves, J., Galli, D., & Walmsley, M. (2004) *A&A* 430:979

Harper, D. A., & Low, F. J. (1971) *ApJ* 165:L9

Harvey, P. M., Gatley, I., Werner, M. W., Elias, J. H., Evans, N. J., II, Zuckerman, B., Morris, G., Sato, T. & Litvak, M. M. (1974) *ApJ* 189:L87

Ho, P. T. P., & Townes, C. H. (1983) *ARA&A* 21:239

Keto, E., Rybicki, G. B., Bergin, E. A. & Plume, R. (2004) *ApJ* 613:355

Kleinmann, D. E., & Low, F. J. (1967) *ApJ* 149:L1

Kramer, C., Richer, J., Mookerjea, B., Alves, J. & Lada, C. J. (2003) *A&A* 399:1073

Lada, C. J., & Kylafis, N. D. (1998) *The Origin of Stars and Planetary Systems* Kluwer Academic Publishers

McKee, C. F. (1989) *ApJ* 345:782

MacLow, M. M. & Klessen, R. S. (2004) *Revs Mod. Phys.* 76:125

Ossenkopf, V. & Henning, T. (1994) *A&A* 291:943

Palla, F. & Stahler, S. W. (2002) *ApJ* 581:1194

Shu, F. H., Adams, F. C. & Lizano,S. (1987) *ARA&A* 25:23.

Solomon, P. M., Sanders, D. B. & Scoville, N. Z. (1979) *ApJ* 232:L89.

Stahler, S. W., Palla, F. (2004) *The formation of stars*, publ. Wiley-VCH

Tafalla, M., Myers, P. C., Caselli, P. & Walmsley, C. M. (2004) *A&A* 416:191

Vastel, C., Phillips, T. G. & Yoshida, H. (2004) *ApJ* 606:L127

Walmsley, C. M., Flower, D. R. & Pineau des Forêts, G. (2004) *A&A* 418:1035

Ward–Thompson,D., André,P. & Kirk, J. M. (2002) *MNRAS* 329:259

Williams, J. P., Bergin, E. A., Caselli, P., Myers, P. C. & Plume, R. (1998) *ApJ* 503:689

Young, C. H., Shirley, Y. L., Evans, N. J. II & Rawlings, J. M. C. (2003) *ApJS* 145:111

First Session in Progress

# IMPULSIVELY TRIGGERED STAR FORMATION

A. P. Whitworth
*School of Physics & Astronomy, Cardiff University, 5 The Parade, Cardiff CF24 3YB, UK*
ant@astro.cf.ac.uk

**Abstract** I review several different modes of impulsively triggered star formation, starting with star formation in turbulent molecular clouds, and exploring the origin of the clump mass function and the scaling relations between clump mass, radius and internal velocity dispersion. This leads to the identification of a critical ram pressure for triggering rapid star formation, and a reappraisal of the minimum mass for opacity-limited star formation. I also discuss star formation triggered by expanding nebulae (HII regions, stellar-wind bubbles and supernova remnants) and star formation triggered by a sudden increase in external pressure, including the formation of brown dwarves and planetary-mass objects in HII regions by photo-erosion of pre-existing prestellar cores. I conclude by describing simulations of interactions between protostellar discs in dense small-N clusters, and note that such interactions are an efficient means of creating low-mass companions.

## Introduction

In this contribution, I shall adopt the viewpoint that many (perhaps most) of the dominant modes of star formation in the Universe are impulsively triggered (e.g. Whitworth et al. 1996), and that star-forming molecular clouds spawn stars on a dynamical time-scale. This picture is sometimes referred to as 'star formation in a crossing time' (Elmegreen 2000). In this picture, molecular clouds are turbulent, because when they form they do not have sufficient time to relax to a quiescent state. Star formation occurs only in those places where two (or more) turbulent flows, having sufficient mass and density, collide with sufficient ram pressure to create a gravitationally unstable dense star-forming core. Therefore, in Section 1, I will review the gravitational fragmentation of a layer formed by the collision of two turbulent flows. In Section 1, I will develop a theory for the origin of the clump mass spectrum, based on gravitational fragmentation. In Section 1, I will derive a critical pressure for impulsively triggered star formation to proceed rapidly, based on the requirement that the shock compressed gas must be able to cool by coupling thermally to the dust. In Section 1, I will revisit the question of opacity limited fragmentation and the minimum mass for star formation, and show that one-shot fragmentation of a shock compressed layer can deliver significantly lower-mass fragments ($\sim 0.003 M_{\odot}$) than conventional hierarchical fragmentation. In Section 1, I will discuss the fragmentation of shells swept up by expanding nebulae, as a means of generating star-forming clumps, and briefly mention a model of Sharples 104 which I am developing with Deharveng and her colleagues. In Section 1, I will describe some work with Hennebelle on the collapse of cores which are subjected to an increase in external pressure, and compare the results with observation. In Section 1, I will present some work with Zinnecker in which we have shown that free-floating brown dwarves and planetary-mass objects can be formed rather easily when cores are overrun by an HII region and stripped down by photo-erosion. Finally, in Section 1, I will describe briefly some simulations of interactions between protostars and protostellar discs, which show that such interactions are an efficient way of generating low-mass companions and free-floating objects.

## Fragmentation of a shock-compressed layer

The gravitational fragmentation of a shock-compressed layer is one of the fundamental microphysical processes in a turbulent, star-forming cloud. The simplest case to consider is where two identical flows of gas, each with density $\rho_{\rm O}$, collide at relative velocity $2v_{\rm O}$. A shocked layer is formed, with isotropic velocity dispersion $\sigma_{\rm S}$ and density $\rho_{\rm S} \sim \rho_{\rm O}(v_{\rm O}/\sigma_{\rm S})^2$. The layer is contained by the ram-pressure $P_{\rm RAM} = \rho_{\rm O} v_{\rm O}^2$ of the material which continues to flow in from either side, *and not by self-gravity*. Consequently, until it fragments, the layer is close to equilibrium, and has a flat density profile. It can fragment while it is still acquiring mass from the two colliding flows.

At their inception, the fastest-growing fragments are shaped like oblate spheroids, with their short dimension $z$ parallel to the inflowing flux of matter, and their long dimension $r$ in the plane of the layer. The time at which the layer starts to fragment non-linearly, and the initial radius, initial thickness and initial mass of the fastest-growing fragments, are given by

$$\begin{aligned} t_{\rm FRAG} &= \left[\frac{\sigma_{\rm S}}{G\,\rho_{\rm O}\,v_{\rm O}}\right]^{1/2}; & r_{\rm FRAG} &= \left[\frac{\sigma_{\rm S}^3}{G\,\rho_{\rm O}\,v_{\rm O}}\right]^{1/2}; \\ m_{\rm FRAG} &= \left[\frac{\sigma_{\rm S}^7}{G^3\,\rho_{\rm O}\,v_{\rm O}}\right]^{1/2}; & z_{\rm FRAG} &= \left[\frac{\sigma_{\rm S}^5}{G\,\rho_{\rm O}\,v_{\rm O}^3}\right]^{1/2} \end{aligned} \quad (1)$$

(Whitworth et al. 1994a,b). It is important to note that these expressions are not the same as those for conventional Jeans fragmentation of an extended 3-dimensional medium having the shocked density, $\rho_{\rm S} \sim \rho_{\rm O}(v_{\rm O}/\sigma_{\rm S})^2$. If we introduce the Mach number of the shock, $\mathcal{M} \equiv v_{\rm O}/\sigma_{\rm S}$, then the initial mass and the initial radius ($m_{\rm FRAG}$, $r_{\rm FRAG}$) are larger by a factor $\mathcal{M}^{1/2}$, and the thickness ($z_{\rm FRAG}$) is smaller by the same factor; hence the ratio of the initial radius to the initial thickness (i.e. the initial edge-on aspect ratio of a fragment) is $r/z \sim \mathcal{M}$.

The equations above have been confirmed by numerical simulations (e.g. Chapman et al. 1992, Pongracic et al. 1992). In general the shock compressed layer at first breaks up into a network of intersecting filaments, and then into prestellar cores along the filaments. The separation between adjacent filaments, and between adjacent cores on a filament, is of order $r_{\rm FRAG}$, as given above. This pattern of fragmentation has the effect that the prestellar cores tend to accrete most of their mass from along the filament out of which they are forming. Since, in general, the two colliding flows are not perfectly aligned, the shock-compressed layer has a net angular momentum, which causes it to tumble. Consequently the filaments in the layer are also tumbling, and as a core accretes from a filament it is spun up until it fragments to produce multiple star systems (Turner et al. 1995, Whitworth et al. 1995).

## Larson's scaling relations and the clump mass spectrum

We can now use the phenomenology of layer fragmentation to derive an explanation for Larson's scaling relations (Larson 1981) and the origin of the clump mass spectrum. In their simplest form, Larson's scaling relations give the radius, $R$, and internal velocity dispersion, $\sigma$, of a clump having mass $M$ – and hence also the mean density, $\rho$, and dynamical time-scale, $t$, of the clump – in terms of simple power laws:

$$\begin{array}{llllll} R(M) & \simeq & R_\bullet\,(M/M_\bullet)^{\alpha}\,; & \rho(M) & \simeq & \rho_\bullet\,(M/M_\bullet)^{1-3\alpha}\,; \\ \sigma(M) & \simeq & \sigma_\bullet\,(M/M_\bullet)^{\beta}\,; & t(M) & \simeq & t_\bullet\,(M/M_\bullet)^{\alpha-\beta}\,, \end{array} \tag{2}$$

where $(R_\bullet, M_\bullet, \rho_\bullet, \sigma_\bullet, t_\bullet)$ are reference values. Similarly, the clump mass spectrum can be approximated by a power law,

$$\begin{array}{lllll} \mathcal{N}_M\,dM & \simeq & K\,M^{-\gamma}\,dM\,, & & \\ \mathcal{M}_M\,dM & \simeq & K\,M^{1-\gamma}\,dM & \equiv & K\,M^{2-\gamma}\,d\ell n[M]\,, \end{array} \tag{3}$$

where $\mathcal{N}_M\,dM$ is the number of clumps having mass in the interval $(M, M + dM)$ and $\mathcal{M}_M\,dM$ is the total amount of mass in clumps having mass in the same interval.

Star formation removes clumps from the low-mass end of this spectrum, but since the overall rate of star formation is slow, the net transfer of mass down the mass spectrum must also be slow. Either the transfer is one-way (from diffuse high-mass clumps to dense low-mass star-forming clumps) and proceeds very slowly at a rate equal to the rate of star formation, or the transfer is two-way and in approximate statistical equilibrium with almost as much mass being transferred up the spectrum by merging as down the spectrum by fragmentation. Either way, the mass spectrum can only be maintained if the mass in equal logarithmic mass bins is proportional to the dynamical time-scale. Combining Eqns. 2 and 3, this yields the approximate constraint:

$$2 - \gamma \simeq \alpha - \beta \qquad \longrightarrow \qquad \gamma \simeq 2 - \alpha + \beta\,. \tag{4}$$

At the high-mass end of the clump mass spectrum, the clumps are supported mainly by turbulence (i.e. the line-widths are non-thermal), and the internal velocity dispersion decreases with decreasing mass. Therefore we can envisage an hierarchy of clumps within clumps, such that a clump of mass $M$ contains subclumps of mass $fM$, and these in turn contain 'sub-sub-clumps' of mass $f^2M$. Assuming that the clumps, sub-clumps and sub-sub-clumps in this hierarchy subscribe to Larson's scaling relations, their properties must adopt the values tabulated below

| | clump | sub-clump | sub-sub-clump |
|---|---|---|---|
| Mass | $M$ | $fM$ | $f^2M$ |
| Velocity dispersion | $\sigma(M)$ | $f^{\beta}\sigma(M)$ | $f^{2\beta}\sigma(M)$ |
| Mean density | $\rho(M)$ | $f^{1-3\alpha}\rho(M)$ | $f^{2-6\alpha}\rho(M)$ |

with $\sigma(M)$ and $\rho(M)$ given by Eqn. (2). Thus, subclumps of mass $fM$ having mean density $f^{1-3\alpha}\rho(M)$ typically collide at velocity $\sigma(M)$ (i.e. the internal velocity dispersion of their parent clump) and the resulting shock-compressed layer typically fragments into subsubclunps of mass $f^2M$ having internal velocity dispersion $f^{2\beta}\sigma(M)$. We can therefore substitute $m_{\rm FRAG} \rightarrow f^2M$, $\sigma_{\rm S} \rightarrow f^{2\beta}\sigma(M)$, $\rho_{\rm O} \rightarrow f^{1-3\alpha}\rho(M)$ and $v_{\rm O} \rightarrow \sigma(M)$ in Eqn. 1, to obtain

$$f^{14\beta+3\alpha-5}\left(\frac{M}{M_\bullet}\right)^{6\beta+3\alpha-3} = \frac{G^3\,M_\bullet^2\,\rho_\bullet}{\sigma_\bullet^6}\,. \tag{5}$$

Since this must be true for any $f$ and any $M$, both the exponents on the lefthand side of Eqn. 5 must be zero, which in combination with Eqn. 4 gives

$$\text{FOR TURBULENT HIGH-MASS CLUMPS:}\qquad \alpha \simeq \frac{1}{2}\,, \qquad \beta \simeq \frac{1}{4}\,, \qquad \gamma \simeq \frac{7}{4} \tag{6}$$

(cf. Williams et al. 1994). Additionally, the righthand side of Eqn. 5 must equal unity, so possible reference values are $M_\bullet = M_\odot$, $\rho_\bullet = 4\times10^{-20}\,{\rm g\,cm^{-3}}$ (corresponding to $n_{\rm H_2} = 10^4\,{\rm H_2\,cm^{-3}}$), $\sigma_\bullet = 0.2\,{\rm km\,s^{-1}}$ (corresponding to the isothermal sound speed in molecular gas at temperature $T = 10\,{\rm K}$),

$R_\bullet = 0.08\,\mathrm{pc}$, and $t_\bullet = 0.4\,\mathrm{Myr}$. (With this choice of reference values, the clumps are self-gravitating, but the analysis leading to Eqn. (6) does not require this assumption.)

At the low-mass end of the clump mass spectrum, the clumps are supported mainly by thermal pressure, and – within a factor of 1.5 – the temperature is $T_\bullet = 10\,\mathrm{K}$, so the isothermal sound speed is $a_\bullet = 0.2\,\mathrm{km\,s^{-1}}$. (The gas cannot readily cool below the cosmic microwave background at $T_{\mathrm{CMB}} \sim 3\,\mathrm{K}$, and efficient cooling by molecular line emission and dust emission keeps the temperature close to $10\,\mathrm{K}$.) Hence, if these low-mass clumps are gravitationally bound, $R \simeq GM/a_\bullet^2 \propto M^1$ and $\sigma \simeq a_\bullet \propto M^0$, giving

$$\text{FOR THERMAL LOW-MASS CLUMPS:} \qquad \alpha \simeq 1, \qquad \beta \simeq 0, \qquad \gamma \simeq 1. \tag{7}$$

Consequently there is a knee in the clump mass function around $M_\bullet \sim M_\odot$, where the switch occurs between the high-mass regime (where core support is dominated by turbulence and $\gamma \simeq 1.75$) and the low-mass regime (where core support is dominated by thermal pressure and $\gamma \simeq 1$). We emphasize that this result presumes an approximately steady statistical equilibrium. If there were a sudden increase in external pressure, this would presumably perturb this equilibrium, driving matter preferentially down to lower masses (and ultimately to protostellar collapse) and thereby steepening the clump mass function somewhat.

## The critical pressure for rapid star formation

A critical issue in the fragmentation of shock compressed layers is the ability of the shocked gas to cool radiatively; indeed, gravitational fragmentation of a shock compressed layer requires better than isothermal cooling across the shock, as first demonstrated by Stone (1970). This requirement cannot be satisfied by line cooling. It requires the full bandwidth afforded when the gas couples thermally to the dust and the dust then emits continuum radiation. Therefore a critical condition for the fragments of a shock-compressed layer to collapse and form stars is that the time-scale for the gas to couple thermally to the dust, $t_{\mathrm{COUPLE}}$, should be less than the freefall time $t_{\mathrm{FF}}$. Assuming that the velocity dispersion in the shock compressed layer is the same as the isothermal sound speed ($\sigma_{\mathrm{S}} \simeq a_{\mathrm{S}}$), and so the shocked density is $\rho_{\mathrm{S}} \simeq \rho_{\mathrm{O}}(v_{\mathrm{O}}/a_{\mathrm{S}})^2$, the time-scale for the gas to couple thermally to the dust is

$$t_{\mathrm{COUPLE}} \simeq \frac{(2\,\pi)^{1/2}\,r_{\mathrm{D}}\,\rho_{\mathrm{D}}}{Z_{\mathrm{D}}\,\rho_{\mathrm{S}}\,a_{\mathrm{S}}} \left( \frac{a_{\mathrm{T}}^2 + a_{\mathrm{S}}^2}{a_{\mathrm{T}}^2} \right). \tag{8}$$

Here $r_{\rm D} \sim 10^{-5}\,{\rm cm}$ is the radius of a representative (spherical) dust grain, $\rho_{\rm D} \sim 3\,{\rm g\,cm^{-3}}$ is the density of the solid material from which dust grains are made, and $Z_{\rm D} \sim 0.01$ is the fraction by mass of dust in the protostellar matter.

The term in brackets on the righthand side of the equation represents the inverse of the thermal accomodation coefficient. This term is approximately unity for $a_{\rm S} < a_{\rm T}$, which with $a_{\rm T} \sim 1\,{\rm km\,s^{-1}}$ corresponds to $T_{\rm S} < 300\,{\rm K}$; for higher temperatures the thermal accommodation coefficient decreases approximately as $T^{-1}$, but we will be mainly concerned with the low temperature regime where thermal accommodation is efficient.

If we now require that $t_{\rm COUPLE} \lesssim t_{\rm FF} = (3\pi/32G\rho_{\rm S})^{1/2}$, we obtain

$$P_{\rm S} \equiv \rho_{\rm S}\,a_{\rm S}^2 \gtrsim P_{\rm COUPLE} \simeq \frac{64\,G}{3}\left[\frac{r_{\rm D}\,\rho_{\rm D}}{Z_{\rm D}}\right]^2 \simeq 10^5\,{\rm cm^{-3}\,K}\,k_{\rm B}\,, \tag{9}$$

where $k_{\rm B}$ is the Boltzmann constant. We emphasize two important aspects of this result. (i) Since in our model of dynamically triggered star formation $P_{\rm S} = \rho_{\rm S} a_{\rm S}^2 = P_{\rm RAM} = \rho_{\rm O} v_{\rm O}^2$, this is in effect a threshold for the ram pressure which will trigger rapid star formation in a turbulent cloud. (ii) $10^5\,{\rm cm^{-3}\,K}\,k_{\rm B}$ is typical of the turbulent pressures found in star-forming molecular clouds; for a self-gravitating cloud, it corresponds to a critical surface-density

$$\Sigma_{\rm COUPLE} \simeq \left(\frac{P_{\rm COUPLE}}{G}\right)^{1/2} \simeq 1.4 \times 10^{-2}\,{\rm g\,cm^{-2}}\,,$$

or equivalently to a critical column-density $N_{\rm H_2} \simeq 4 \times 10^{21}\,{\rm H_2\,cm^{-2}}$.

## The minimum mass for star formation

The minimum mass for star formation is normally attributed to opacity limited 3-dimensional hierarchical fragmentation, as first described by Hoyle (1953). In this picture, a protocluster cloud contracts under gravity, and as long as the gas can cool efficiently – i.e. as long as the gas remains isothermal – the Jeans mass decreases with increasing density and so the cloud can break up into sub-clouds, and the sub-clouds can break up into sub-sub-clouds, and so on, until the initial cloud is broken up into many small prestellar cores. This process of hierarchical fragmentation halts rather suddenly when the fragments become so dense and opaque that they can no longer cool efficiently, the temperature rises, the Jeans mass no longer decreases with increasing density, and fragmentation then ceases. An elegant analysis by Rees (1976) shows that the minimum mass for 3-dimensional hierarchical fragmentation is given by

$$M_{\rm MIN} \simeq \frac{m_{\rm PLANCK}^3}{\bar{m}^2}\left(\frac{a}{c}\right)^{1/2} \propto T^{1/4}\,\bar{m}^{-9/4}\,, \tag{10}$$

where $m_{\rm PLANCK} \equiv (hc/G)^{1/2} \simeq 5.5 \times 10^{-5}\,{\rm g}$ is the Planck mass, $\bar{m} \simeq 4 \times 10^{-24}\,{\rm g}$ is the mean gas-particle mass, $a$ is the isothermal sound speed, and $c$ is the speed of light. A more detailed analysis by Silk (1977) gives a value of order $0.015 M_\odot$ for contemporary (Population I) star formation in the solar vicinity ($Z \simeq Z_\odot$).

However, there are problems with this picture. Hierarchical fragmentation is never observed in numerical simulations, and there is no evidence for its occuring in nature. This probably has to do with the following factors. (i) The time-scale on which a subcloud contracts is always longer than the time-scale on which its parent cloud is contracting. Therefore neighbouring subclouds tend to be merged by the overall contraction of the parent cloud, before they can condense out as separate entities. Permanent fragmentation is therefore unlikely, unless nonlinear perturbations are applied at the outset (i.e. 'turbulence' is invoked, e.g. Goodwin et al. 2004a, 2004b), but even then the fragmentation is not noticeably hierarchical. (ii) Isolated subclouds are likely to grow by accretion as they contract. Thus, even if a subcloud starts out with a mass just exceeding the local Jeans mass, its mass will have increased by a very large factor before its density contrast with the background becomes significant.

Given these problems with hierarchical fragmentation, and in the light of the hypothesis that star formation regions may be very dynamical, Boyd & Whitworth (2004) have considered whether the opacity limit gives a different minimum mass when applied to one-shot 2-dimensional fragmentation of a shock-compressed layer. As in Section 1, we consider a shock compressed layer formed by the collision of two idenitical streams having pre-shock density $\rho_{\rm O}$ and relative velocity $2v_{\rm O}$. We again assume that the velocity dispersion in the shock compressed layer is the same as the isothermal sound speed ($\sigma_{\rm S} \simeq a_{\rm S}$), and the unperturbed shocked density is therefore $\rho_{\rm S} \simeq \rho_{\rm O}(v_{\rm O}/a_{\rm S})^2$. Protofragments are modelled as spheroids with initial half-thickness $z_{\rm I}$ equal to the half-thickness of the layer. The initial radius $r_{\rm I}$ is chosen by identifying the fastest growing fragment. Fragmentation occurs whilst the layer continues to grow, and the opacity limit is set by the requirement that the heating rate,

$$\mathcal{H} \;\simeq\; -P\,\frac{dV}{dt} + 2\pi r^2 \rho_{\rm O} v_{\rm O} \left( v_{\rm O}^2 - \dot{r}^2 \right), \qquad (11)$$

must not exceed the maximum possible radiative cooling rate,

$$\mathcal{C} \;\simeq\; 2\pi r^2 \sigma_{\rm SB} T_{\rm S}^4 \,. \qquad (12)$$

The first term on the righthand side of Eqn. 11 is the rate at which the internal energy of the fragment is increased by compressional work. The second term on the righthand side is the rate at which energy is dissipated in the shock where the inflowing material is decelerated, minus the energy used to accelerate this

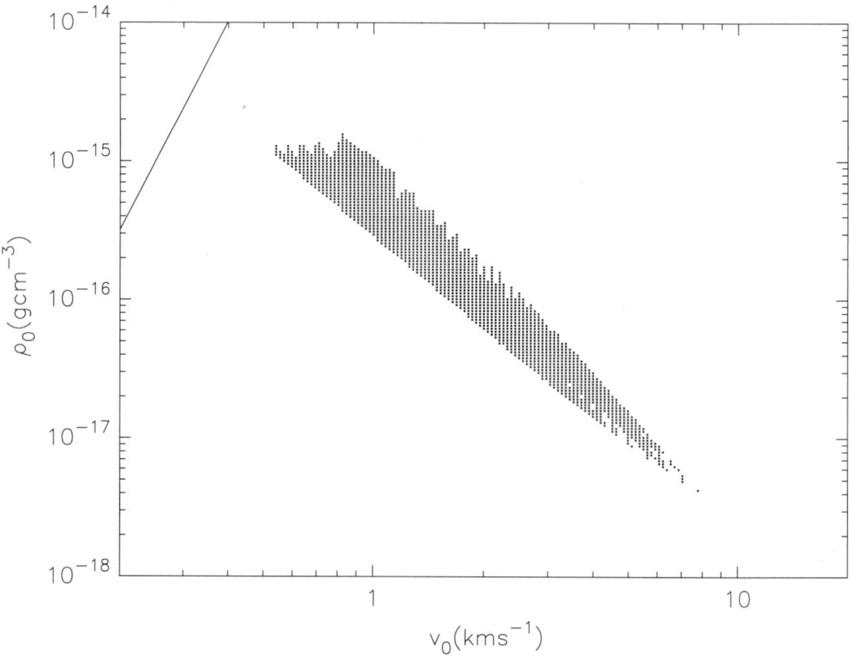


*Figure 1.* A log/log plot of the $(\rho_\text{O}, v_\text{O})$ plane. The dots mark combinations of pre-shock density, $\rho_\text{O}$, and collision speed, $v_\text{O}$, for which the fastest-growing fragment mass is less than $0.005M_\odot$. We assume an effective post-shock sound speed $a_\text{S} = 0.2\,\text{km}\,\text{s}^{-1}$, corresponding to molecular gas at 10 K. The irregularities in the boundary of this region have to do with the tendency for low-mass fragments to pulsate before collapsing; one extra pulsation can make the difference between collapsing with $M_\text{FINAL} < 0.005\,\text{M}_\odot$ or accumulating more mass and hence ending up with $M_\text{FINAL} > 0.005\,\text{M}_\odot$. The solid line is the locus below which $\rho_\text{O}$ must fall if the immediate post-shock cooling radiation is to be treated in the way we have assumed.

material up to the speed with which the fragment is condensing laterally. Typically the first and second terms are comparable. Adopting $a_\text{S} = 0.2\,\text{km}\,\text{s}^{-1}$ (corresponding to molecular gas, $\bar{m} \simeq 4 \times 10^{-24}\,\text{g}$, at $T_\text{S} = 10\,\text{K}$), we find

$$M_\text{MIN} \simeq 0.0027M_\odot\,. \tag{13}$$

Figure 1 shows the region of the $(\rho_\text{O}, v_\text{O})$-plane in which the minimum mass is less than $0.005M_\odot$. We see that there is a wide range of pre-shock densities ($5 \times 10^{-18}\,\text{g}\,\text{cm}^{-3} \lesssim \rho_\text{O} \lesssim 2 \times 10^{-15}\,\text{g}\,\text{cm}^{-3}$) and shock velocities ($0.6\,\text{km}\,\text{s}^{-1} \lesssim v_\text{O} \lesssim 6\,\text{km}\,\text{s}^{-1}$) over which very low mass fragments can be created by the fragmentation of shock-compressed layers.

We note that this is a conservative estimate for the minimum mass, since the immediate post-shock gas is much hotter than $T_\text{S} = 10\,\text{K}$, and therefore the energy dissipated in the accretion shock is radiated at higher temperature – hence more efficiently – than the $PdV$ work. We also note that this is a more realistic derivation of the minimum mass than the standard derivation of the minimum mass for 3-dimensional hierarchical fragmentation, on two important counts. (i) It takes account of competition with other scales of fragmentation. Here the minimum mass is required to be the fastest condensing mass-scale. (It is a well-known property of layer fragmentation that there is a finite fragmentation scale which condenses out faster than all others (e.g. Larson 1985).) (ii) The analysis takes account of continuing accretion during condensation. For example the minimum mass fragment with $M_{\rm FINAL} = 0.0027 M_{\odot}$ starts out with $M_{\rm INITIAL} = 0.0011 M_{\odot}$ (Boyd & Whitworth 2004).

## The fragmentation of shells swept up by expanding nebulae

There are other channels for impulsively triggered star formation. For example, on large scales star formation occurs in shells swept up by expanding nebulae, i.e. expanding HII regions, stellar wind bubbles and supernova remnants. This is the basis of sequential self-propagating star formation. If an HII region, stellar wind bubble or supernova remnant is excited by a large cluster of recently-formed stars, it can lead to a burst of star formation of such violence that the interstellar medium is restructured on scales comparable with the host galaxy. In such cases it is necessary to include the effect of galactic rotation (Jungwiert & Palous, 1994). However, the basic mechanism of fragmentation is essentially the same as for layer fragmentation.

As an example, we consider a stellar wind bubble driven by a star (or stars) whose wind is steady and isotropic and has mechanical luminosity $L_{\rm WIND} \equiv \dot{M}_{\rm WIND} v_{\rm WIND}^2 / 2$, where $\dot{M}_{\rm WIND}$ is the mass loss rate and $v_{\rm WIND}$ is the speed of the wind. We suppose that the ambient interstellar medium is static with uniform density $\rho_{\rm O}$ and that this gas is swept up into a thin dense shell at the edge of the bubble; the swept-up gas in the shell has isothermal sound speed $a_{\rm S}$. The spherically symmetric expansion of the bubble can then be approximated by a similarity solution $R_{\rm SHELL} \sim (L_{\rm WIND} t^3 / \rho_{\rm O})^{1/5}$. Non-linear fragmentation of the shell occurs at time $t_{\rm FRAG}$ when the shell has radius $R_{\rm FRAG}$, where

$$t_{\rm FRAG} \simeq \left[ \frac{a_{\rm S}^5}{G^5 \rho_{\rm O}^4 L_{\rm WIND}} \right]^{1/8}, \qquad R_{\rm FRAG} \simeq \left[ \frac{L_{\rm WIND}\, a_{\rm S}^3}{G^3 \rho_{\rm O}^4} \right]^{1/8}. \tag{14}$$

The mass and initial radius of the fastest condensing fragments are then

$$m_{\rm FRAG} \simeq \left[ \frac{a_{\rm S}^{29}}{G^{13} \rho_{\rm O}^4 L_{\rm WIND}} \right]^{1/8}, \qquad r_{\rm FRAG} \simeq \left[ \frac{a_{\rm S}^{13}}{G^5 \rho_{\rm O}^4 L_{\rm WIND}} \right]^{1/8} \tag{15}$$

(Whitworth & Francis 2002). We note that these results are not very sensitive to the assumption of a uniform background density. In other words, they still hold if there is a density gradient in the ambient interstellar medium so that only a partial shell forms. They are also evidently not very sensitive to $L_{\rm WIND}$, except that we must have

$$L_{\rm WIND} > 3 \times 10^4 \frac{a_{\rm S}^5}{G} \simeq 10^3 \, L_\odot \left( \frac{a_{\rm S}}{\rm km\,s^{-1}} \right)^5 ; \qquad (16)$$

otherwise the expansion of the shell stalls, due to the gravity of the accumulated mass, before it fragments. The shell can still fragment after it stalls, but the fragments are then comparable in size to the radius of the shell, and their mass is essentially the same as the Jeans mass in the undisturbed background medium.

With Deharveng and Lefloch I am currently trying to model Sharples 104. This is a spherically symmetric HII region with four molecular fragments distributed evenly around its rim; one of the fragments contains an embedded young star cluster (Deharveng et al. 2003). In our model, a star cluster forms at the centre of a molecular cloud and excites an HII region which then expands, sweeping up a dense shell of gas which eventually breaks up into the four molecular fragments we see today. The model is constrained to fit the observed output of ionizing photons ($\mathcal{N}_{\rm LyC}$), the mass and diameter of the HII region ($M_{\rm HII}$, $d_{\rm HII}$), the mass of molecular gas in the four fragments, the fact that there are just four fragments, and the lack of molecular gas on the line of sight through the centre of the HII region (Whitworth et al. 2004).

## Pre-existing cores subjected to a sudden increase in external pressure

Another type of impulsively triggered star formation occurs when a pre-existing core is subjected to a sudden increase in external pressure. In the simplest case, we consider a stable Bonnor-Ebert sphere, contained by an external pressure $P_{\rm EXT}$, and then we increase $P_{\rm EXT}$ on a time-scale shorter than, or comparable with, the sound-crossing time (Hennebelle et al. 2003). This has the effect of driving a compression wave into the core. The compression wave leaves in its wake a subsonic and approximately uniform inward velocity field, similar to what is inferred from the asymmetric double-peaked molecular-line profiles seen in L1544 and other prestellar cores (e.g. Tafalla et al. 1998, Williams et al. 1999, Lee et al. 1999, 2001, Gregersen et al. 2000). When the compression wave reaches the centre, it is reflected as an expansion wave, and a protostar is formed. The protostar grows by accretion, at first quite rapidly ($\gtrsim 10^{-5} M_\odot\,{\rm yr}^{-1}$) and then less rapidly ($\lesssim 10^{-6} M_\odot\,{\rm yr}^{-1}$). This pattern of initially rapid and then decreasing accretion accords well with observations of Class 0 and Class I sources. (Class 0 sources are relatively rare, and their outflows are powerful, implying a short-lived period of rapid accretion. Class I sources are more frequent and have weaker outflows, implying a longer-lived period of slower accretion.)

In these spherically symmetric simulations, only a single central protostar is formed. However, if a little rotation is added (Hennebelle et al. 2004), the result is usually a binary or triple system. For $\beta \lesssim 0.05$ (where $\beta$ is the initial ratio of rotational to gravitational energy), the large-scale flows are the same as in the non-rotating case and a central primary protostar forms, but then a disc forms around the central primary protostar and the disc fragments to form secondary protostars. The disc tends to be more unstable – in the sense that it spawns more secondary protostars – if $\beta$ is larger or the rate of compression is higher. This is because, as $\beta$ is increased, more matter is deposited in the outer parts of the circumprimary disc; and as the rate of compression is increased, the matter is deposited in the outer parts of the disc more rapidly and at higher density. All these factors contribute to making the disc more unstable against gravitational fragmentation. For very high $\beta$ and very rapid compression, there is no central primary protostar. Instead fragmentation occurs via the formation of a ring.

## Forming free-floating brown dwarves and planetary-mass objects by photo-erosion

One way in which the pressure acting on a pre-existing core might be increased is if the surrounding gas is suddenly ionized. With Zinnecker (Whitworth & Zinnecker 2004) I have developed a semi-analytic model for a pre-existing core which is overrun by an HII region. We have shown that this could be a very effective way of forming free-floating brown dwarves and/or planetary-mass objects. As soon as the surroundings of the core are ionized, the increase in external pressure drives a compression wave into the core. When this compression wave impinges on the centre of the core, a protostar forms there and starts to grow by accretion. At the same time, the compression wave is reflected as an expansion wave and propagates outwards, leaving in its wake an approximately freefall velocity field, which feeds the accretion flow onto the growing central protostar. Whilst all this is going on, an ionization front eats into the core, eroding its outer layers. The final mass of the central protostar is determined by the moment at which the inward propagating ionization front encounters the outward propagating expansion wave. Soon after that moment, the kinetic energy of the newly ionized gas flowing off the ionization front becomes less than its binding energy to the central protostar, and photo-erosion ceases. To a good approximation the final protostellar mass is given by

$$M_{\rm FINAL} \simeq 0.01\mathrm{M}_\odot \left(\frac{a_{\rm I}}{\rm km\,s^{-1}}\right)^6 \left(\frac{\dot{\mathcal{N}}_{\rm LyC}}{10^{50}\,\rm s^{-1}}\right)^{-1/3} \left(\frac{n_{\rm O}}{10^3\,\rm cm^{-3}}\right)^{-1/3}, \quad (17)$$

where $a_{\rm I}$ is the isothermal sound speed in the neutral gas of the core, $\dot{\mathcal{N}}_{\rm LyC}$ is the rate at which the central OB star (or stars) emit hydrogen-ionizing photons, and $n_\text{O}$ is the density in the ambient HII region. There is evidently a wide range of parameter space in which free-floating brown dwarves and planetary-mass objects can be formed by this process – provided there are suitable cores pre-existing in the vicinity of an OB star when it switches on its ionizing output. However, although the mechanism is very effective – in the sense that it works for a wide range of likely conditions – it is also very inefficient, – in the sense that it takes a rather massive pre-existing core to form a single brown dwarf or planetary-mass object.

## Forming low-mass companions and free-floating objects in interactions between protostellar discs

Another mechanism for forming low-mass objects impulsively involves violent interactions between protostellar discs. Most protostars are surrounded by accretion discs when they first form, and initially these discs are likely to be quite massive (i.e. disc mass comparable to the mass of the central protostar) and quite extended (radius $\gtrsim 100\,\text{AU}$). At the same time, it appears that about 40% of stars are born in binaries with separations less than $\sim 200\,\text{AU}$; and numerical simulations suggest that collapsing cores fragment when they reach the density $\rho_\text{CRIT} \sim 10^{-13}\,\text{cm}^{-3}$ at which the gas switches from approximate isothermality to approximate adiabaticity, i.e. fragmentation occurs on scales $\sim (M_\odot/\rho_\text{CRIT})^{1/3} \sim 100\,\text{AU}$. Therefore it seems inevitable that protostellar discs will experience violent interactions with other protostellar discs, and with naked protostars. Indeed, such interactions are probably an essential element in the birth of a small-$N$ subcluster from a star-forming core.

With Watkins, Boffin Bhattal and Francis (Boffin et al. 1998, Watkins et al. 1998a,b) I have simulated a large ensemble of such interactions. In order to treat the gas as isothermal, we consider a $1M_\odot$ disc with radius $1000\,\text{AU}$ around a $1M_\odot$ protostar. We limit the approach orbit to the parabolic case, but we sample all the different possibilities for the relative orientation of the spins and the orbit, and many different periastra. In general we find interactions to be very efficient in triggering disc fragmentation, and hence the formation of additional low-mass protostars. More violent encounters (i.e. encounters with smaller periastra) are more effective in this regard, but even distant 'tidal' encounters lead to fragmentation. Encounters in which the spin and orbital angular momenta are aligned are also somewhat more effective, but even with random alignments of the spins relative to the orbit the mechanism still appears to be effective in triggering disc fragmentation. On average, an encounter leads to the formation of $\sim 2.4$ additional protostars if the spins are aligned with the orbit, and $\sim 1.2$ additional protostars if the spins are random. Most of these new protostars remain bound to the initial protostars, and in this case they acquire their own discs and continue to grow by accretion – as long as there is a supply of accretable material and they can compete effectively for it. The rest of the newly-formed protostars are ejected dynamically, and these are normally discless (i.e. low-mass Weak-Lined T Tauri stars).

The critical parameter in these simulations is the viscosity, $\eta$. If we use the disc viscosity parametrization of Shakura & Sunyaev (1973; $\eta \simeq \alpha_{\rm SS} a^2/\Omega$, where $a$ is the sound speed and $\Omega$ is the angular rotation speed of the disc), the masses of the additional protostars are concentrated in the range $0.010 - 0.100 M_\odot$ (i.e. brown dwarves) for $\alpha_{\rm SS} \simeq 10^{-2}$, and in the range $0.001 - 0.010 M_\odot$ (i.e. planetary-mass objects) for $\alpha_{\rm SS} \simeq 10^{-3}$. In view of the brown-dwarf desert, and the relative abundance of exo-planets, this might be interpreted as evidence for the lower value of $\alpha_{\rm SS} \simeq 10^{-3}$; but this is very speculative.

## Conclusions

We have shown that there are many different possible modes of impulsively triggered star formation, and that these modes seem likely to make a significant contribution to the star formation in the Universe. Impulsively triggered modes of star formation have a number of advantages over more spontaneous – and therefore more quasistatic – modes (for instance star formation which is regulated by the loss of magnetic flux through ambipolar diffusion). In particular, (i) impulsive triggers have the potential to co-ordinate star formation across a large region of space, and thereby to explain the approximate co-evality of star formation in clusters. (ii) By launching matter directly into the non-linear regime of instability against fragmentation, impulsive triggers promote the formation of multiple systems, and of multiple systems having a wide range of parameters (mass ratios, separations and eccentricities). (iii) Impulsive triggering would seem to be an essential element of sequential self-propagating star formation and of starbursts, but it is probably also important even in more quiescent star formation regions like Taurus (e.g. Hartmann et al. 2002). (iv) Impulsive triggering can produce free-floating low-mass objects. In contrast, spontaneous, quiescent modes of star formation have difficulty explaining the approximatre co-evality of star formation across extended star clusters, the frequency of multiple star systems and their wide range of parameters, and the time-scales for the different phases of star formation (as inferred from the statistics of the different classes of prestellar core and protostar).

## Acknowledgments

I gratefully acknowledge the support of the EC for the Research Training Network on *The Formation and Evolution of Young Stellar Clusters* (HPRN-CT-2000-00155), and many collaborations, in particular those with with Henri

Boffin, Douglas Boyd, Lise Deharveng, Simon Goodwin, Patrick Hennebelle, Dimitri Stamatellos and Hans Zinnecker.

## References


Boffin, H. M. J., Watkins, S. J., Bhattal, A. S., Francis, N. & Whitworth, A. P. (1998). *MNRAS*, 300:1189.

Boyd, D. F. A. & Whitworth, A. P. (2004). *A&A*, in press

Chapman, S. J., Pongracic, H., Disney, M. J., Nelson, A. H., Turner, J. A. & Whitworth, A. P. (1992). *Nature* 359:207

Deharveng, L., Lefloch, B., Zavagno, A., Caplan, J., Whitworth, A. P., Nadeau, D. & Martin, S. (2003). *A&A* 408:L25

Deharveng, L., Lefloch, B., Zavagno, A., Caplan, J., Whitworth, A. P., Nadeau, D. & Martin, S. (2004) in preparation

Elmegreen, B. G. (2000).*ApJ* 530:277

Goodwin, S. P., Ward-Thompson, D. & Whitworth A. P. (2004a).*A&A* 414:633

Goodwin, S. P., Ward-Thompson, D. & Whitworth, A. P. (2004b).*A&A* 419:543

Gregersen, E. M., Evans, N. J. II, Mardones, D. & Myers, P. C. (2000).*ApJ* 533:440

Hartmann, L., Ballesteros-Paredes, J. & Bergin, E. A. (2002).*ApJ* 562:852

Hennebelle, P., Whitworth, A. P., Gladwin, P. P. & André, P. (2003).*MNRAS* 340:670

Hennebelle, P., Whitworth, A. P., Cha, S. H. & Goodwin, S. P. (2004).*MNRAS* 348:687

Hoyle, F. (1953).*ApJ* 118:513

Jungwiert, B. & Palous, J. (1994).*A&A* 287:55

Larson, R. B. (1981).*MNRAS* 194:809

Larson, R. B. (1985).*MNRAS* 214:379

Lee, C. W., Myers, P. C. & Tafalla, M. (1999).*ApJ* 526:788

Lee, C. W., Myers, P. C. & Tafalla, M. (2001).*ApJS* 136:703

Pongracic, H., Chapman, S. J., Davies, J. R., Disney, M. J., Nelson, A. H. & Whitworth, A. P. (1992).*MNRAS* 256:291

Rees, M. J. (1976).*MNRAS* 176:483

Silk, J. I. (1977).*ApJ* 214:152

Stone, M. E. (1970).*ApJ* 159:293

Stutzki J. Güsten, R. (1990).*ApJ* 356:513

Tafalla, M., Mardones, D., Myers, P. C., Caselli, P., Bachiller, R. & Benson, P. J. (1998).*ApJ* 504:900

Turner, J. A., Chapman, S. J., Bhattal, A. S., Disney, M. J., Pongracic, H. & Whitworth, A. P. (1995).*MNRAS* 277:705

Watkins, S. J., Bhattal, A. S., Boffin, H. M. J., Francis, N. & Whitworth, A. P. (1998a).*MNRAS* 300:1205

Watkins, S. J., Bhattal, A. S., Boffin, H. M. J., Francis, N. & Whitworth, A. P. (1998b).*MNRAS* 300:1214

Whitworth, A. P., Bhattal, A. S., Chapman, S. J., Disney, M. J. & Turner, J. A. (1994a).*A&A* 290:421

Whitworth, A. P., Bhattal, A. S., Chapman, S. J., Disney, M. J. & Turner, J. A. (1994b).*MNRAS* 268:291

Whitworth, A. P., Bhattal, A. S., Francis, N. & Watkins, S. J. (1996).*MNRAS* 283:1061

Whitworth, A. P., Chapman, S. J., Bhattal, A. S., Disney, M. J., Pongracic, H. & Turner, J. A. (1995).*MNRAS* 277:727

Whitworth, A. P., Deharveng, L., Lefloch, B., Zavagno, A., Caplan, J., Nadeau, D. & Martin, S. (2004). in preparation

Whitworth, A. P. & Francis, N. (2002).*MNRAS* 329:641

Whitworth, A. P. & Zinnecker, H. (2004).*A&A* 427:299

Williams, J. P., Myers, P. C., Wilner, D. J. & Di Francesco, J. (1999).*ApJ* 513:L61

Williams, J. P., de Geus, E. J. & Blitz, L. (1994).*ApJ* 428:693

Paola Caselli speaking on astro chemistry

Carlos Eiroa and Andre Moitinho.

# STARLESS CORES

Mario Tafalla
*Observatorio Astronómico Nacional, Alfonso XII 3, E-28014 Madrid, Spain*
m.tafalla@oan.es

**Abstract** Dense low mass cores in nearby clouds like Taurus and Auriga are some of the simplest sites currently forming stars like our Sun. Because of their simplicity and proximity, dense cores offer the clearest view of the different phases of star formation, in particular the conditions prior to the onset of gravitational collapse. Thanks to the combined analysis of the emission from molecular lines and the emission/absorption from dust grains, the last several years have seen a very rapid progress in our understanding of the structure and chemical composition of starless cores. Previous contradictions between molecular tracers are now understood to arise from core chemical inhomogeneities, which are caused by the selective freeze out of molecules onto cold dust grains. The analysis of the dust emission and absorption, in addition, has allowed us to derive accurate density profiles, and has made finally possible to carry out self consistent modeling of the internal structure of starless cores. In this paper I briefly review the evolution of core studies previous to the current golden age, and show how multi-tracer emission can now be modeled in a systematic manner. Finally I show how we can start to reconstruct the early history of core formation taking advantage of the chemical changes in the gas.

## Introduction

Dense starless cores in nearby clouds like Taurus and Auriga represent the simplest star-forming sites. They collapse and produce individual stars (or binaries) in almost isolation, with apparently little influence from the surrounding cloud or from previous generations of stars. For this simplicity and proximity, starless cores constitute ideal places to elucidate the still mysterious process by which interstellar matter collapses and forms gravitationally-bound self-luminous objects.

The simplicity of star-formation in cores, unfortunately, comes at the price of missing some elements of interest for this meeting, like clustering or the formation of high-mass stars. Several reviews in this book provide information on these topics, and the reader is referred to them to complement the material presented here. Despite the above limitations, however, protostars formed in starless cores display most phenomena that we associate with the birth of a typical star. Gravitational motions, binary and disk formation, and bipolar outflow ejection all occur in young stellar objects (YSOs) recently formed (or in the process of formation) in low mass cores, and most of these phenomena were in fact first identified in dense core environments. Thus, if we are to understand the basic principles and time evolution of most star-formation physics, studying dense cores offers a most promising approach.

When we observe a starless core, we are seeing a system that very likely will collapse to form a star (it is still an open question whether all starless cores are truly pre stellar). For this reason starless cores offer a snapshot of the initial conditions of star formation, and even of the first stages of the process if they have already started to collapse. Deriving these conditions from observations is therefore of great interest, and more than two decades of dense core studies attest the enormous effort made in this direction. For a number of reasons discussed below, the last five years or so have witnessed a very rapid progress in the study of dense cores, and numerous results and papers have appeared in the literature. In this contribution I review some of this recent work, with emphasis on the study of the internal structure of dense cores. In the pedagogical spirit of the workshop, I start with a brief review of the field. I have chosen a historical point of view (my own simplistic summary of the history) to emphasize what we have learned, what are some of the most relevant papers in the field, and where should be aiming to go in the future. Contributions in this book by Malcolm Walmsley and Paola Caselli treat related topics from different points of view and are highly recommended.

## "Classical" studies of starless cores

The systematic study of low mass dense cores has a tradition of more than two decades, and starts in earnest with the first searches based on the optical inspection of Palomar Survey plates by Myers et al. (1983) (see Lee & Myers 1999 for an update of this work). These searches were followed by radio studies of their molecular emission, mostly $NH_3$ (e.g., Myers & Benson 1983), to determine the basic gas physical properties. Subsequent correlation of dense core positions and the location of IRAS point sources by Beichman et al. (1986) confirmed that dense cores are the sites of ongoing star formation. IRAS and NIR observations, in addition, gave rise to a division of cores into star-containing and starless, depending on whether they harbored or not a central luminous object. It should be emphasized, however, that this classical distinction is based on a given level of sensitivity achieved more than a decade ago, and estimated by Myers et al. (1987) as corresponding to a luminosity of 0.1 $L_\odot$ for the distance of Taurus. The recent revolution in mid IR sensitivity brought by the Spitzer Space Telescope has forced some corrections in the old classification of cores, and moved some cores previously classified as starless to the star-containing category (Young et al. 2004). The distinction between starless and star-containing cores, however, is still a fundamental one, in the sense that it distinguishes between before and after the formation of a central object.

The 1980s represent the golden years of mapping and analysis of dense cores based on their molecular line emission, what we could call the "classical period". The main results from this decade are summarized in the monumental paper by Benson & Myers (1989), whose 15th birthday closely coincides with the celebration of this meeting in October 2004. Benson & Myers's work is based on arcminute-resolution observations of a single tracer ($NH_3$), and concentrates on the global or average properties of cores. After observing and analyzing more than 100 dense cores (both starless and star-containing), these authors derive, among other properties, typical masses of a few $M_{\odot}$, sizes of about 0.1 pc, densities of a few $10^4$ cm$^{-3}$, temperatures of 10 K, and subsonic turbulent motions. To recognize the importance of these results one only needs to notice that these observational parameters constituted part of the input for the so-called standard model of star formation, which crystallized almost at the same time (Shu, Adams, & Lizano 1987).

The above set of dense core properties was not free from contradictions, as already discussed in the almost simultaneous paper by Zhou et al. (1989). These authors compared observations of cores in two high density tracers, CS and $NH_3$, and found systematic discrepancies between them. The most severe discrepancies were a mismatch in the position of the emission peaks of the two molecules, a factor of 2 larger size of the CS maps, and a factor of 2 broader CS lines. Although different mechanisms were invoked at the time to explain these CS/$NH_3$ discrepancies, the lack of a detailed picture of the internal structure of dense cores left the problem unsolved. As we will see below, only now, after more than a decade of work by a large number of researchers, we can claim a reasonable understanding of the internal properties of starless cores and finally put an end to the old contradiction between different tracers.

## Starless cores studies in the 1990s

The 1990s witnessed an enormous progress in the characterization of dense cores in general and starless cores in particular. This progress resulted from a decade of improvements in the techniques to trace the gas and dust components of dense cores, and in the parallel development of sophisticated tools to analyze the data. The classical method of molecular-line tracing greatly benefited from the factor of several increase in spatial resolution afforded by the new generation of millimeter and submillimeter telescopes like the IRAM 30m and the JCMT. It also benefited from the advent of multipixel receiver arrays, like QUARRY and SEQUOIA at FCRAO, which have allowed the systematic mapping of large areas at moderate resolution. As a result, it is now possible to map and resolve large numbers of dense cores in relatively short amounts of time (e.g., Caselli et al. 2002).

More important than the incremental increase in our ability to trace the molecular emission of cores has been the development of new techniques to sample the dust component. This component can be traced by its millimeter and submillimeter thermal emission, which is optically thin and most likely arises from dust of close to constant emission properties. Pioneer work using mm and submm continuum observations has provided the first realistic density profiles of starless cores, and has shown that they present a characteristic flattening at their center (Ward-Thompson et al. 1994, Andre et al. 1996). Such a flattening implies that the dust (and gas) density in the inner 5,000-10,000 AU of a core has a close-to-constant value which typically ranges between $10^5$ to $10^6$ $cm^{-3}$ (Ward-Thompson et al. 1999).

Additional progress in the tracing of the dust component comes from the technique of NIR extinction measurement, which uses the reddening of background stars behind a core or a cloud to estimate the dust column density (Lada et al. 1994). Using this technique, Alves et al. (2001) has produced an extremely detailed view of the B68 globule, and has shown that its density profile is very close to that expected for a close-to-critical isothermal (Bonnor-Ebert) sphere. A related technique, which uses the extinction by a core of the mid-IR diffuse background, has been used by Bacmann et al. (2000) to complement the mm and submm continuum emission measurements by estimating density profiles and confirming the presence of central flattening.

When the above dust measurements and molecular line observations were combined to carry out a unified analysis of starless cores, it became clear that many molecules do not sample the central and densest part of the cores, but that they disappear from the gas phase most likely due to their freeze out onto cold dust grains (Kuiper et al. 1996, Kramer et al. 1999, Caselli et al. 1999, Tafalla et al. 2002). This freeze out of molecules onto grains is not only of chemical interest by itself (see Bergin & Langer 1997 for pioneer modeling), but has important consequences for dense core studies. Molecular emission is the only tool available to study gas kinematics, so the disappearance of some species at high densities imposes a significant limitation in this type of studies. As a positive aspect, the presence of selective freeze out of molecules in the central gas of dense cores helps to explain the contradictions between CS and $NH_3$ observations described by Zhou et al. (1989), because these two molecules behave differently at high densities: CS freezes out easily and $NH_3$ does not (see Tafalla et al. 2002 for further details).

The 1990s was also a decade of intense studies of dense core kinematics, spurred by the identification of infall motions in the star-containing core B335 (Zhou et al. 1993). Among starless cores, L1544 was the first one found to present evidence for inward motions (Myers et al. 1996, Tafalla et al. 1998), and subsequent searches by Lee et al. (1999) have identified several additional cases. These initial infall searches were made using molecular species like CS and $H_2CO$, now known to freeze out at densities typical of dense core centers, so they were only sensitive to motions in the outer layers of the cores. More recent infall studies take molecular freeze out into consideration, and can therefore penetrate deeper in the gas.

## The internal structure of the l1498 and L1517B cores

To show the level of detail currently possible in the analysis of the internal structure of starless cores, this section summarizes the recent study of two cores in the Taurus-Auriga complex, L1498 and L1517B. These two cores (Fig. 1) were selected for their close-to-round shape and relative isolation, in addition to being starless (no IRAS or 2MASS sources), and they have been subject to observations in the 1.2mm continuum and a number of molecular lines using the FCRAO 14m, IRAM 30m and Effelsberg 100m telescopes (Fig. 2). A complete account of is this work is presented in Tafalla et al. (2004; 2005, in preparation).

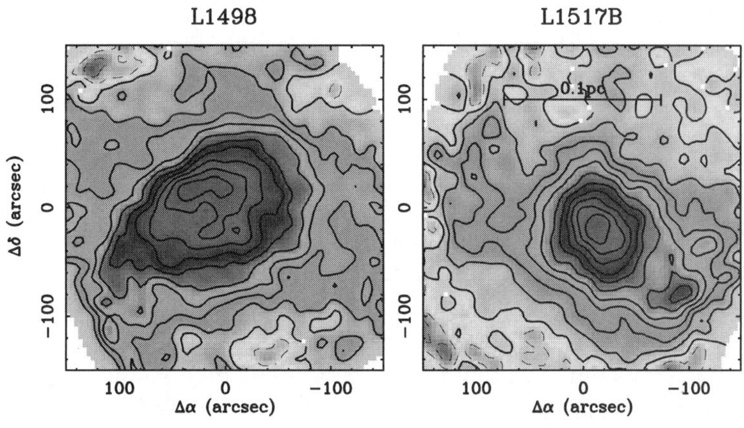


*Figure 1.* 1.2mm continuum maps of L1498 and L1517B. Note their central concentration and close-to-round geometries. First contour and contour spacing are 2 mJy/11″-beam.

Characterizing the internal structure of a dense core implies deriving both its physical parameters (density, temperature, turbulence) and its chemical properties (molecular abundances) as a function of position. Given that we only see the cores projected on the plane of the sky, we need to make an assump-

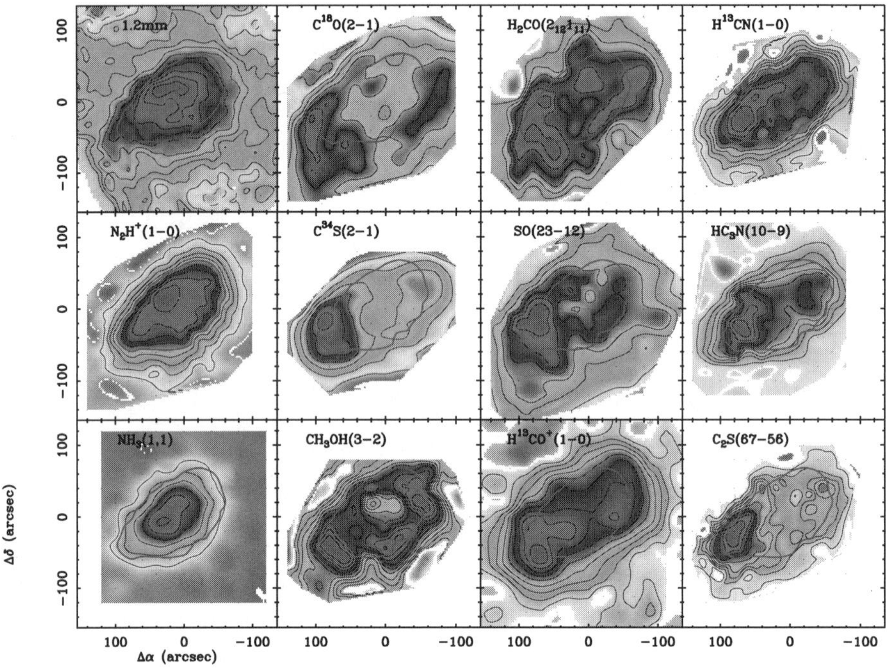


*Figure 2.* Sample of molecular emission maps of the L1498 core. Only the maps in the first column show centrally peaked emission, while the rest present a dip indicative of molecular depletion. For easier inter comparison, the half maximum contour of the $N_2H^+$(1–0) emission has been superposed to each map (red line in the color version of the figure).

tion on their third (line-of-sight) dimension, and for the close-to-round targets of our study, we will assume spherical symmetry. Once we have fixed the geometry, we can use the optically thin 1.2mm continuum emission to derive a density profile, and the observations of appropriate molecular lines to derive the gas temperature, turbulence, and systemic motions in addition to the abundance profiles of each species. Because of the gradient in gas excitation caused by the density gradient, a non-LTE radiative transfer solution is necessary to model the molecular emission. For this work, we use the Monte Carlo radiative transfer code of Bernes (1979) updated with the most recent molecular parameters and collision rates. This model is spherically symmetric, so we compare its results with radial averages of the emission, and fit simultaneously the radial profile of integrated intensity and the central spectrum for each line. The following subsections show the analysis step by step and the conclusions derived from it.

## Density

The first step in our analysis consists of deriving the density profile of each core based on its 1.2mm continuum emission. Unfortunately, this step is the most uncertain one, due to our incomplete knowledge of the physical properties of the emitting dust. In the optically thin case (a good approximation at 1.2mm), the dust emission depends on the product of the dust emissivity and the Planck function at the dust temperature, and both emissivity and dust temperature are not well constrained. For this work, we assume that the two parameters have constant values of 0.005 cm $g^{-1}$ and 10 K, respectively. Our choice of dust temperature is based on our estimate of the gas temperature using $NH_3$ and presented in the next section, while our choice of emissivity follows the standard value in the literature (e.g., Andre et al. 1996). If grain growth via coagulation is important toward the core center, the true emissivity can increase there, and be larger than our assumed value by a factor of about two (Ossenkopf & Henning 1994). The dust temperature, on the other hand, may decrease toward the center due to the extinction of the interstellar radiation field (the dominant heating mechanism for dust in core environments), so it may reach a lower value of around 8 K for our cores (Evans et al. 2001, Zucconi et al. 2001, Galli et al. 2002). Thus, if for example we use an emissivity of 0.009 cm $g^{-1}$, as recommended by Ossenkopf & Henning (1994) (their OH5 value), and a dust temperature of 8.7 K, as derived by Galli et al. (2002) for a core similar to L1517B, we will have to decrease our densities by a factor 0.7. Our chosen values, therefore represent a simplification of the complex dust physics, but their product seems more immune to this physics that each of the individual components.

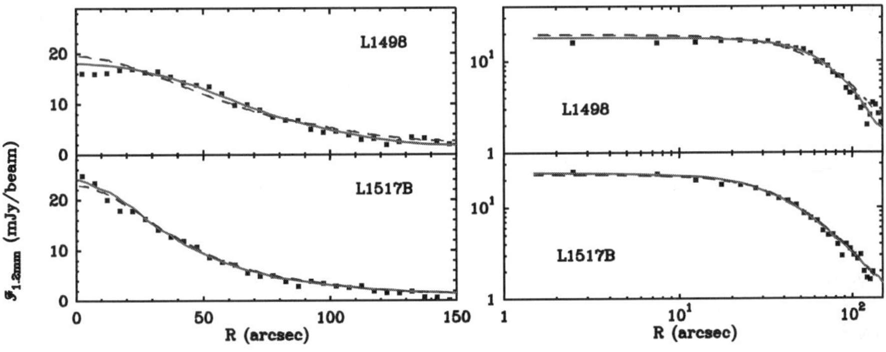


*Figure 3.* Radial profiles of 1.2mm continuum emission in linear-linear and log-log scales (left and right, respectively). The squares represent bolometer observations, the solid lines are the predictions from an analytic model, and the dashed line are the predictions from an isothermal model (see text).

Once we have chosen the dust parameters, we can invert the observed emission into a density profile. To do this, we start with a choice of density distribution and predict its expected emission by solving the equation of radiative transfer and simulating an on-the-fly observation with a bolometer array (including secondary chopping). The result of this prediction is compared with the observed radial profile, and the density distribution is corrected accordingly; the process is repeated until the model matches the observations. As density profiles, we have used two families of curves. One is the empirical formula

$$n(r) = n_0/(1 + (r/r_0)^\alpha),$$

where $n_0$, $r_0$, and $\alpha$ are free parameters that correspond to the central density, half maximum radius, and asymptotic power law of the density profile. Note that this profile contains naturally the central flattening first found by Ward-Thompson et al. (1994). The other family of density profiles is the family of isothermal (Bonnor-Ebert) profiles, which has only two free parameters, the central density and the effective sound speed.

The result from the fitting procedure is shown in Figure 3, where both linear-linear and log-log radial profiles of the observed 1.2mm intensity (squares) are shown together with the predictions from the analytic (solid lines) and isothermal (dashed-lines) models. As can be seen, both families of models fit similarly well the data, with the analytic model for L1498 fitting slightly better the profile than the isothermal model. Both families of models require similar central densities, $10^5$ cm$^{-3}$ for L1498 and 2 $10^5$ cm$^{-3}$ for L1517B, and the size of the half maximum radius (from the analytic model) is approximately 10,000 AU for L1498 and 5,000 AU for L1517B. L1498, therefore, is slightly less dense and more extended than L1517B, but both cores have similar central column densities: 3-4 $\times 10^{22}$ cm$^{-2}$ (30-40 Av). With respect to the $\alpha$ parameter, we derive 3.5 and 2.5 for L1498 and L1517B, respectively. The 2.5 value for L1517B makes the analytic fit indistinguishable from isothermal fit, as it can be shown that the analytic formula becomes an excellent approximation to the Bonnor-Ebert sphere for the case of $\alpha = 2.5$ (see Tafalla et al. 2004, Appendix B).

## Temperature

To derive the gas temperature profiles of L1498 and L1517B, we use $NH_3$(1,1) and (2,2) observations. The combined analysis of these two lines can provide a reliable and straightforward estimate of the gas kinetic temperature with minimal assumptions, and to achieve that, we have updated the classical $NH_3$ analysis of Walmsley & Ungeretchs (1983) using a Monte Carlo radiative transfer solution. The derived radial profiles of gas kinetic temperature are shown in

Figure 4 (left panel), and for both cores are relatively flat with a constant value of 10 K.

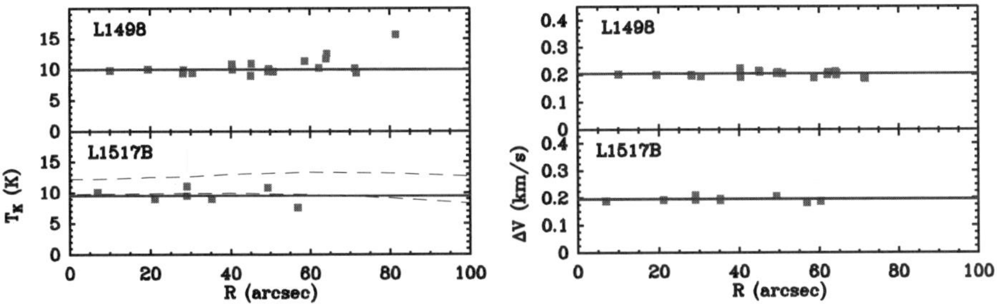


*Figure 4.* *Left:* radial profiles of gas kinetic temperature as derived from the analysis of the $NH_3$ emission. The solid lines are constant temperature fits at 10 K and the dashed lines are the predictions from two models by Galli et al. (2002). *Right:* radial profiles of total $NH_3$ linewidth as derived from a hyperfine analysis to the (1,1) spectrum. The lines represent two constant linewidth fits with $\Delta V = 0.2$ km s$^{-1}$.

At the central gas densities of L1498 and L1517B, dust and gas are expected to be thermally coupled, and therefore follow similar gradients with radius. As mentioned before, the dust kinetic temperature is expected to drop towards the core center (Evans et al. 2001, Zucconi et al. 2001), but our gas temperature profiles do not show such a trend. Our $NH_3$ observations have a 40″ resolution (FWHM), so it is still possible to hide a small temperature gradient, but not a very large one given the strong weight of the $NH_3$ emission towards the core center due to its higher abundance there (see below). To test our sensitivity to a temperature gradient, we have used the realistic temperature profiles predicted by Galli et al. (2002) for a core very similar to L1517B, and simulated an $NH_3$ observation with 40″ resolution followed by a temperature analysis. The results, shown as dashed lines in Fig. 4, indicate that to fit the data, a low cosmic rate is needed (1/5 of standard), and that for this case, which has a central drop of about 1 K, the fit is as good as the constant temperature fit. The central gas temperature in L1517B, therefore, may be slightly lower than in the outside, but not by much more than 1 K. Such a strong constrain arises from the strong sensitivity to temperature of the $NH_3$(2,2), whose intensity would double by just increasing the temperature by 3 K.

## Turbulence

The $NH_3$ lines also provide a sensitive measure of the gas turbulent motions via the hyperfine analysis of the (1,1) spectra. Figure 4 (right panel) presents the radial profiles of total $NH_3$ linewidth (FWHM) for L1498 and L1517B, which are again flat with a constant value of 0.2 km s$^{-1}$ and a very small dispersion ($\leq$ 0.01 km s$^{-1}$, consistent with random errors). This constant linewidth with radius is in contrast with the so-called linewidth-size relation (e.g., Larson 1981) in two ways. First, the linewidth does not increase with radius, and second, the linewidth of the more extended L1498 core is the same as that of the more compact L1517B. The lack of linewidth increase with radius in dense cores has been discussed in detail by Goodman et al. (1998).

The total linewidth represented in Figure 4 is the harmonic average of a thermal and a non-thermal components. For $NH_3$ molecules at 10 K, the thermal component is 0.16 km s$^{-1}$, so this component dominates the linewidth, while the turbulent component is subthermal (see also Myers 1983). To compare the contribution of the two components to the equilibrium of the core, we take the ratio between thermal and non-thermal pressures:

$$\frac{P_{NT}}{P_T} = \frac{\sigma_{NT}^2}{kT/m} \approx 0.07,$$

where $\sigma_{NT}$ is the turbulent component ($= \Delta V/\sqrt{8\ ln2} = 0.05$ km s$^{-1}$) and $m$ is the mean particle mass (2.33 the proton mass). As the ratio indicates, the contribution of the turbulent component to the support of the core is negligible in the central 0.1 pc.

## Molecular composition

Once the density, temperature and kinematics of the cores have been determined, the only free parameter we can adjust to model the observed emission of any molecule is the abundance profile of that species (see Tafalla et al. 2004 for a discussion on the velocity fields in L1498 and L1517B, omitted here for lack of space). This means that we can use the L1498 and L1517B cores as laboratories of known physical properties to measure the behavior of molecules at high densities and determine their sensitivity to freeze out onto dust grains. In order to simplify the modeling and the comparison between molecules, we use for all species an abundance profile with a constant value of $X_0$ at large radius and a step central hole at $R_{\rm hole}$ ($NH_3$ has a central increase). For each species at least two transitions have been observed, and for each of them we model simultaneously both its radial profile of integrated intensity and its central emerging spectrum. This is done with a Monte Carlo radiative transfer code (Bernes 1979) together with a convolution routine to simulate an astronomical observation.

A summary of the modeling results is presented in Figure 5, where the radial profiles for one transition of each species in L1498 are shown (squares) together with the best fit model (solid line) and a model with constant abundance equal to the outer value of the best fit model (dashed lines). As can be seen, all molecules but $N_2H^+$ and $NH_3$ require a central abundance hole of

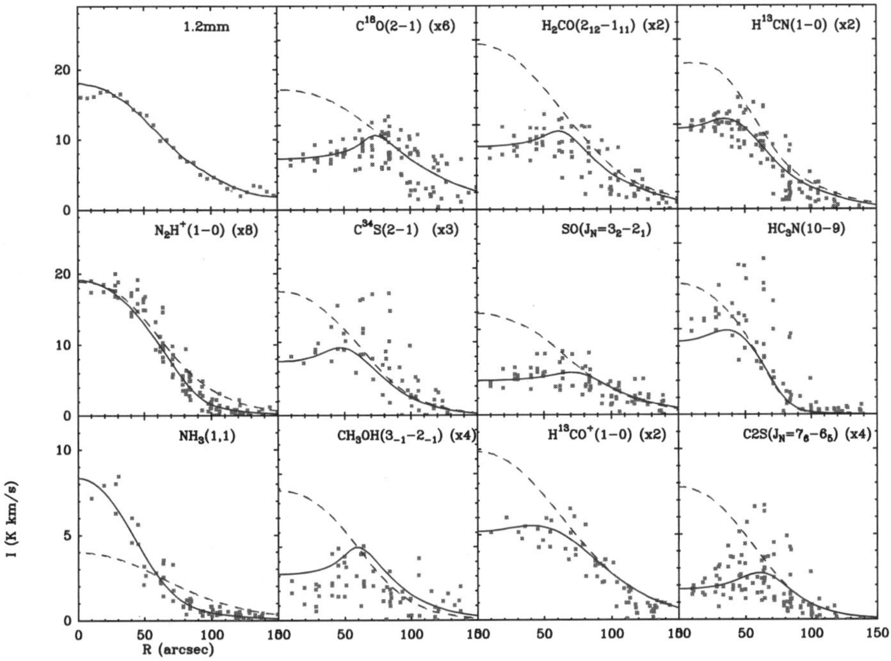

*Figure 5.* Radial profiles of emission from L1498 for the tracers shown in Figure 2. The squares represent observations, the dashed lines are the predictions from constant abundance models set to fit the outer emission, and the solid lines are the best fit models (they also fit additional lines not shown here). For all species in the second, third, and fourth columns, the best fit models contain a central abundance hole.

variable size (5,000-10,000 AU), which indicates that they freeze out at densities of a few $10^4$ cm$^{-3}$. A similar behavior is found in L1517B.

The above pattern of abundances in L1498 and L1517B is in general agreement with recent models of starless core chemistry, which show that only species based on the simplest nitrogen chemistry can survive for a significant time at densities of $10^5$ cm$^{-3}$ (e.g., Bergin & Langer 1997, Shematovich et al. 2003, Aikawa et al. 2003). This behavior results from a combination of the lower binding energy of the $N_2$ molecule to dust grains, which allows N chemistry to proceed at high densities, together with the favorable chemistry resulting from the disappearance of CO from the gas phase (see chapter by Paola Caselli in this volume for more details). Although the general agreement between models and observations is good, there are still some puzzling behaviors, like the abundance increase of $NH_3$ at high densities while the abundance of its cousin species $N_2H^+$ stays constant. Further modeling of the high density core chemistry is still needed to explain the data (see Aikawa et al. 2005).

From the point of view of observations, the results from L1498 and L1517B are a strong reminder that very few molecules survive in the gas phase at the typical densities of a starless core. Extreme care should therefore be exercised when choosing a molecular tracer to study the properties of the star-forming material in these environments.

## Tracing core evolution with molecular freeze out

Our new understanding of molecular freeze out is finally allowing a self consistent picture of core interiors to emerge. It also promises to shed crucial light on the still mysterious process of core contraction. This is so because molecular freeze out is a progressive and irreversible process at the densities and temperatures of starless cores. Each molecule that hits a dust grain under these conditions will stick to its surface and will not evaporate thermally (cosmic ray-induced evaporation seems not efficient enough to reverse the process, see Leger et al. 1985, Hasegawa & Herbst 1993). The amount of freeze out in a given parcel of gas, therefore, will increase with time, and it should be possible in principle to convert the amount of freeze out into an estimate of the contraction age. In practice, unfortunately, this is not yet achievable because of the uncertainties in the binding energies of molecules to dust grains (but see Aikawa et al. 2005 for a recent detailed modeling). Molecular freeze out, however, can already be used as a qualitative time marker thanks to the simple fact that it increases with time. This means that by looking at the amount of freeze out of a sensitive molecule (like CO or CS), we can distinguish cores at different contraction stages, although we cannot yet assign them a particular age.

Molecular freeze out is not the only process that increases with time in a contracting core. Some molecules like $NH_3$ and $N_2H^+$ gradually become more abundant because their formation depends on the slow formation of $N_2$ (e.g., Suzuki et al. 1992), so they are also useful indicators of core evolution ($N_2H^+$ is further enhanced by the depletion of CO). Thus, we can expect that a core starts its life being CO/CS rich at its center (not enough time to freeze out) and $NH_3/N_2H^+$ poor (not enough time to form $N_2$) and evolves toward being centrally CO/CS poor (due to freeze out) and $NH_3/N_2H^+$ rich (due to late-time chemistry). To quantify this trend, we define a ratio between the intensities of two molecules with opposite behaviors. For observational convenience, we chose the two thin species $C^{18}O$ and $N_2H^+$, as they have J=1-0 transitions at 3mm, and can therefore be observed with the same telescope at similar resolutions. Thus, we define

$$R = I[\mathrm{C}^{18}\mathrm{O}(1-0)] \;/\; I[\mathrm{N_2H^+}(1-0)],$$

where both integrated intensities are measured at the core center (as defined by the mm continuum peak). A young core is expected to have strong $C^{18}O$

emission and weak $N_2H^+$ emission, so it will be characterized be a relative large value of $R$. Conversely, an evolved core will be weak in $C^{18}O$ and bright in $N_2H^+$, so its $R$ ratio will be low. Using typical abundances of $C^{18}O$ and $N_2H^+$ in dense cores (Tafalla et al. 2004), we estimate that the dividing line between young and evolved cores occurs at about $R = 1$.

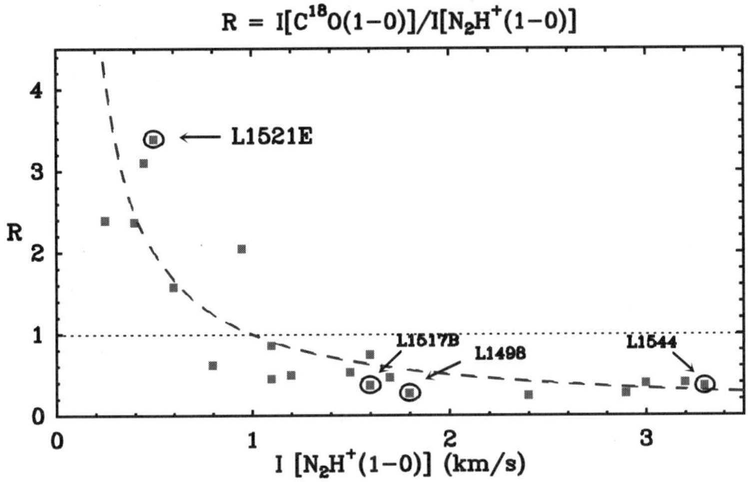


*Figure 6.* Plot of the $R$ factor ($R = I[C^{18}O(1-0)]/I[N_2H^+(1-0)]$) as a function of $N_2H^+$(1–0) intensity for a sample of starless cores. This factor seems to be an indicator of core evolution, and to decrease as a core ages. The positions of the extremely young core L1521E, the intermediate-age cores L1498 and L1517B, and the highly evolved L1544 are indicated. The dashed line is the prediction from a simplified model of core evolution.

Figure 6 presents a plot of the $R$ ratio as a function of $N_2H^+$(1–0) central intensity for a sample of 21 starless cores. Most cores lie below the $R = 1$ line, suggesting that $C^{18}O$ is frozen out at their center, a fact confirmed by Monte Carlo radiative transfer calculations for some of them (L1498, L1517B, L1544, see Tafalla et al. 2002). As these cores were selected from $NH_3$ surveys (like that of Benson & Myers 1989), the presence of freeze out at their center is an indication that they belong to a population of objects already chemically evolved, and therefore relatively "old" (of course, not old enough to have formed a star).

In order to find cores with a ratio $R > 1$ indicative of chemical youth, we have selected starless cores known to have weak or undetected $NH_3$ emission, and we have mapped them in $C^{18}O$ and $N_2H^+$ with the 14m FCRAO telescope (Tafalla et al. 2005, in preparation). As a result of this search, we have identified a small population of starless cores with $R > 1$, which at the same time have a relatively low value of $I[N_2H^+(1-0)]$. These cores populate the upper-left region of the diagram in Fig. 6, and represent the best candidates for chemically young starless cores. The object with largest $R$ value is L1521E in Taurus, and because of its extreme characteristics, it has been the subject of a more detailed study.

## L1521E: the youngest starless core?

To understand the origin of the extreme $R$ value in L1521E, we have observed this core with high resolution using the IRAM 30m telescope (see Tafalla & Santiago 2004 for a full report). These observations have been made in the 1.2mm continuum, $N_2H^+$(1–0), and several rare isotopes of CO, and the results have been analyzed with the same method as the L1498 and L1517B data presented in section 4. A gas and dust kinetic temperature of 10 K has been assumed, as suggested by the observations of (very weak) $NH_3$ lines with the Effelsberg 100m telescope.

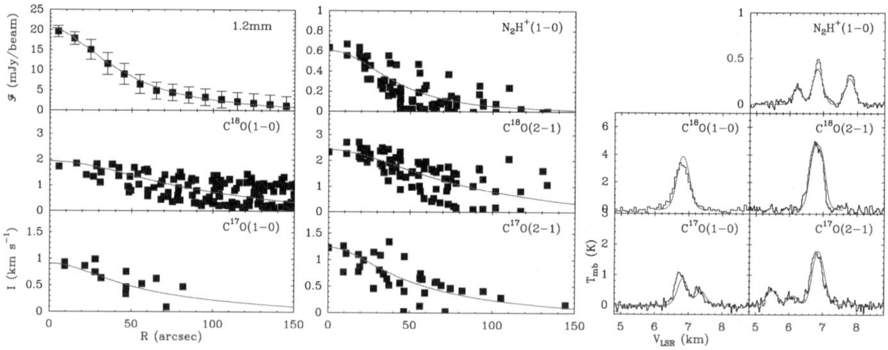


*Figure 7.* Radial profiles of integrated intensity (left) and emerging spectra (right) of different tracers towards L1521E. The squares and histograms are observed data, and the lines are model predictions. In contrast with the results from the L1498 analysis (Figure 5), the CO isotopomer emission from L1521E can be fit with a constant abundance model. This makes L1521E the first starless cores with negligible CO depletion.

The main results of the L1521E analysis are summarized in Fig. 7, which shows radial profiles of integrated intensity and emerging central spectra for all our data. As in the L1498 and L1517B analysis, the squares represent the observed data and the lines are fits from a Monte Carlo radiative transfer calculation (or an optically thin analysis for the 1.2mm continuum). In contrast with the L1498 and L1517B cores, however, all CO isotopomer data from L1521E can be fit assuming a constant abundance, together with a $C^{18}O/C^{17}O$ isotopic ratio of 3.65 (Penzias 1981). In addition, the value of the $C^{18}O$ abundance in L1521E is almost the same as the value needed to fit the outer part of the L1498 and L1517B cores. This all suggests that $C^{18}O$ is not frozen out at the center of L1521E, or if it is, it happens only in a very small region diluted by our resolution and sampling (about 20″). In this respect, L1521E is the first starless core consistent with no freeze out.

The very large $R$ value of L1521E ($R = 3.4$) not only arises from a lack of significant $C^{18}O$ depletion. It also results from a very low value of its $N_2H^+$ abundance, which according to the Monte Carlo calculations is 8 times lower than that estimated for L1498 and L1517B. As mentioned before, a low $N_2H^+$ abundance is another indication of extreme youth given the late-time character of this species. More evidence for a L1521E being a young core comes from previous work by Suzuki et al. (1992) and Hirota et al. (2002), who also concluded that L1521E is unusually young based on its very bright CS and $C_2S$ emission and its weak $NH_3$ lines. It is reassuring that all these different lines of work consistently indicate that the L1521E core is unusually chemically unprocessed.

Given the multiple indications of chemical youth, it is surprising that the density profile of L1521E is very similar to that of more evolved cores like L1517B. At least naively, one would have expected that a young core should have a lower central density and a more distended (cloud-like) structure, but clearly this is not the case for L1521E. Whether this is unique to this core (as a the result of fast contraction) or typical of all young cores requires the study of more objects that lie above the $R = 1$ line in Figure 6. Work currently in progress will hopefully answer this question soon.

## Acknowledgments

It is a pleasure to thank the organizers of this workshop, especially Nanda Kumar, for an enjoyable and productive meeting. Part of the work presented here is the result of an ongoing collaboration with Joaquin Santiago, Phil Myers, Paola Caselli, Malcolm Walmsley, and Claudia Comito. I thank them warmly for help and many discussions on starless cores and related issues over the last several years.

## References

Aikawa, Y., Ohashi, N., Herbst, E. 2003, ApJ, 593, 906

Aikawa, Y., Herbst, E., Roberts, H., & Caselli, P. 2005, ApJ, 620, 330

Alves, J.F., Lada, C.J., & Lada, E.A. 2001, Nature, 409, 159

Andre, P., Ward-Thompson, D., & Motte, F. 1996, A&A, 314, 625

Bacmann, A., Andre, P., Puget, J.-L., Abergel, A., Bontemps, S., & Ward-Thompson, 2000, A&A, 361, 558

Beichman, C.A., Myers, P.C., Emerson, J.P., Harris, S., Mathieu, R., Benson, P.J., & Jennings, R.E. 1986, ApJ, 307, 337

Benson, P.J. & Myers, P.C. 1989, ApJS, 71, 89

Bergin, E.A. & Langer, W.D. 1997, ApJ, 486, 316

Bernes, C. 1979, A&A, 73, 67

Caselli, P., Benson, P.J., Myers, P.C., Tafalla, M., 2002, ApJ, 572, 238

Caselli, P., Walmsley, C.M., Tafalla, M., Dore, L., & Myers, P.C. 1999, ApJ, 523, L165
Evans, N.J.,II., Rawlings, J.M.C., Shirley, Y.L., Mundy, L.G. 2001, ApJ, 557, 193
Galli, D., Walmsley, C.M., & Gon calves, J. 2002, A&A, 394, 275
Goodman, A.A., Barranco, J., Wilner, D.J., & Heyer, M.H. 1998, ApJ, 504, 223
Hasegawa, T.I. & Herbst, E. 1993, MNRAS, 261, 83
Hirota, T., Ito, T., & Yamamoto, S. 2002, ApJ, 565, 359
Kramer, C., Alves, J., Lada, C.J., Lada, E.A., Sievers, A., Ungerechts, H., Walmsley, C.M. 1999, A&A, 342, 257
Kuiper, T.B.H., Langer, W.D., Velusamy, T. 1996, ApJ, 468, 761
Lada, C.J., Lada, E.A., Clemens, D.P., & Bally, J. 1994, ApJ, 429, 694
Larson, R.B. 1981, MNRAS, 194, 809
Leger, A., Jura, M., & Omont, A. 1985, A&A, 144, 147
Lee, C.W., & Myers, P.C. 1999, ApJS, 123, 233
Lee, C.W., Myers, P.C. & Tafalla, M. 1999, ApJ, 526, 788
Myers, P.C. 1983, ApJ, 270, 105
Myers, P.C., Bachiller, R., Caselli, P., Fuller, G.A., Mardones, D., Tafalla, M., Wilner, D.J. 1996, ApJ, 449, L65
Myers, P.C., & Benson, P.J. 1983, ApJ, 266, 309
Myers, P.C., Linke, R.A., & Benson, P.J. 1983, ApJ, 264, 517
Myers, P.C., Fuller, G.A., Mathieu, R.D., Beichman, C.A., Benson, P.J., Schild, R.E., Emerson, J.P. 1987, ApJ, 319, 340
Ossenkopf, V. & Henning, T. 1994, A&A, 291, 943
Penzias, A.A. 1981, ApJ, 249, 518
Shematovich, V.I., Wiebe, D.S., Shustov, B.M., Li, Z.-Y. 2003, ApJ, 588, 894
Shu, F.H., Adams, F.C., & Lizano, S. 1987, ARA&A, 25, 23
Suzuki, H., Yamamoto, S., Ohishi, M., Kaifu, N., Ishikawa, S.-I., Hirara, Y., & Takano, S. 1992, ApJ, 392, 551
Tafalla, M., Mardones, D., Myers, P.C., Caselli, P., Bachiller, R., & Benson, P.J. 1998, ApJ, 504, 900
Tafalla, M., Myers, P.C., Caselli, P., Walmsley, C.M., & Comito, C. 2002, ApJ, 569, 815
Tafalla, M., Myers, P.C., Caselli, P., & Walmsley, C.M. 2004, A&A, 416, 191
Tafalla, M. & Santiago, J. 2004, A&A, 423, L21
Walmsley, C.M. & Ungeretchs, H. 1983, A&A, 122, 164
Ward-Thompson, D., Motte, F., & Andre, P. 1999, MNRAS, 305, 143
Ward-Thompson, D., Scott, P.F., Hills, R.E., Andre, P. 1994, MNRAS, 268, 276
Young, C.H. et al. 2004, ApJS, 154, 396
Zhou, S., Evans, N.J., II, Kömpe, C., Walmsley, C.M. 1993, ApJ, 404, 232
Zhou, S., Wu, Y., Evans, N.J.,II, Fuller, G.A., & Myers, P.C. 1989, ApJ, 346, 167
Zucconi, A., Walmsley, C.M., & Galli, D. 2001, A&A, 376, 650

# CHEMICAL PROCESSES IN STAR FORMING REGIONS

Paola Caselli
*INAF–Osservatorio Astrofisico di Arcetri, Largo E. Fermi 5, I–50125 Firenze, Italy*
caselli@arcetri.astro.it

This paper will review the basic concepts of gas–phase and grain surface chemistry of dense molecular clouds, where low mass and high mass stars form. The chemistry of cold pre-stellar cloud cores, where molecular freeze-out and deuterium fractionation dominate, will be presented. Then, following cloud evolution after protostellar birth, hot core and shock chemistry will be discussed in view of recent observations. A brief summary of the chemistry in protoplanetary disks will also be furnished. The aim is to identify important gas tracers in the various steps of the star formation process, pointing out the main problems still open in the field of astrochemistry.

## Introduction

A good comprehension of the physical properties of molecular clouds and star forming regions implies a detailed radiative transfer analysis of molecular line data and their interpretation with chemical models, inclusive of gas–phase reactions as well as gas–grain and grain surface processes. In fact, molecular lines allow us to determine volume densities and temperatures (e.g. Evans 1999), independently from dust continuum emission and absorption measurements (e.g. Ward–Thompson et al. 1999; Bacmann et al. 2000; Visser et al. 2002). This is important to refine the results derived from the dust continuum observations, which are affected by (factor of a few) uncertainties due to the poorly known properties of dust grains (e.g. coagulation degree, thickness of icy mantles; Ossenkopf & Henning 1994; see also Bianchi et al. 2003) and the presence of dust temperature gradients (e.g. Evans et al. 2001; Galli et al. 2002; Pagani et al. 2004). Furthermore, molecular lines offer a unique way to study the kinematics of dense cores (e.g. Ohashi et al. 1999; Lee et al. 2001; Shinnaga et al. 2004).

The number of known interstellar molecules is continuously growing (see http://astrochemistry.net for a constantly updated list of interstellar molecules). Up to now, 137 molecules have been detected in the interstellar medium (205 including isotopomers) and about 50 in comets. The largest species so far observed contains 13 atoms ($HC_{11}N$; Bell & Matthews 1985). The present paper will not be an exhaustive review of the whole field of astrochemistry. It has the aim of introducing the basic concepts of gas phase and surface chemistry (Sect. 7) and showing what are the main chemical processes in star forming regions (Sect. 1), from dark clouds to protoplanetary disks. The hope is to highlight the beauty of this field, which has become a crucial ingredient in all the areas of astrophysics. Interested readers are invited to see the following reviews: van Dishoeck & Blake (1998), Evans (1999), Ehrenfreund & Charnley (2000), van Dishoeck (2004). Useful books are "The chemically controlled cosmos" by Hartquist & Williams (1995); "Interstellar Chemistry" by Duley & Williams (1984); "The formation of stars" by Stahler & Palla (2004); "The Physics and Chemistry of the Interstellar Medium" by Tielens (2005).

## Basic concepts in astrochemistry

The ingredients of astrochemical models are: atoms (with elemental abundances with respect to hydrogen nuclei reported in Tab. 1), molecular species, electrons, dust grains, cosmic rays and UV photons. Models of dark clouds typically neglect the action of UV photons (which is a good assumption when the visual extinction is larger than $\sim 4$ mag; e.g. McKee 1989), and start with all the elements in atomic form (although sometimes hydrogen is assumed to be in molecular form initially; e.g. Leung et al. 1984).

The first problem that chemists have to face concerns the elemental abundances. Four different values are listed in Tab. 1 for each atomic species, two measured and two typically adopted in chemical models: (i) the *solar* abundance, measured in the solar photosphere (from Anders & Grevesse 1989, but see Asplund et al. 2005 for recent revisions on these values); (ii) the *cosmic* abundance, slightly but significantly (factor between 1.5 and 2) lower than the solar values, measured in the direction of stars in the solar neighborhood (Snow & Witt 1996); (iii) the *high metal* abundances, used for diffuse cloud models, in which elements are depleted as found in diffuse clouds according to Morton (1975); (iv) the *low metal* abundances, used for dense cloud chemistry, where the elements S, Si, Na, Mg, and Fe are further depleted by two orders of magnitude. The *low metal* mix has been used to best reproduce the observed abundances of sulfur–containing species (Prasad & Huntress 1982), complex molecules (Herbst & Leung 1989), and the ionization balance of interstellar clouds (Graedel et al. 1982).

A better way to proceed would be to start with diffuse–cloud conditions and follow elemental depletion with time until dark cloud conditions are reached; however, this implies a knowledge of ($i$) the contraction time scale, which is still debatable (e.g. Palla & Stahler 2002; Hartmann 2003), ($ii$) the probability for each element to stick onto a dust grain (the so–called sticking coefficient,

*Table 1.* Measured elemental abundances (with respect to total hydrogen) and those adopted in chemical models.

| Element | MEASURED Solar | Cosmic | ADOPTED High Metal | Low Metal |
|---|---|---|---|---|
| C | 3.6(-4) | 2.1(-4) | 7.3(-5) | 7.3(-5) |
| N | 9.3(-5) | 6.6(-5) | 2.1(-5) | 2.1(-5) |
| O | 7.4(-4) | 4.6(-4) | 1.8(-4) | 1.8(-4) |
| S | 1.9(-5) | 1.2(-5) | 8.0(-6) | 8.0(-8) |
| Si | 3.6(-5) | 1.9(-5) | 8.0(-7) | 8.0(-9) |
| Fe | 3.2(-5) | 2.7(-5) | 3.0(-7) | 3.0(-9) |
| Na | 1.9(-6) | ... | 2.0(-7) | 2.0(-9) |
| Mg | 3.8(-5) | 2.5(-5) | 7.0(-7) | 7.0(-9) |

Note: a(b) $\equiv$ a$\times 10^b$.

which is typically assumed of order unity, but see Jones & Williams 1985); (iii) the fraction of charged grains and their interaction with positive ions (i.e. all those elements with ionization potential less than that of H in diffuse clouds, such as C, S, Si, Fe, Na, Mg; e.g. Ruffle et al. 1999; Weingartner & Draine 2001).

Once the set of initial abundances has been assumed, one needs to solve the "rate equations", which govern the time variation of species both in the gas–phase and on the surface of dust grains (see e.g. Hasegawa et al. 1992). This involves quite often the assumption of uncertain rate coefficients for gas–phase reactions and the poorly known processes happening on grain surfaces. Nevertheless, chemical models are now reasonably able to reproduce observational data, thanks to the tight interaction between theorists, observers and experimentalists, fundamental for constraining models on the one side and guiding observations on the other. The next two subsections will treat gas–phase and surface chemistry separately, pointing out the basic concepts in both fields.

## Gas–phase chemistry

The rate coefficient $k$(cm$^3$/s) of a generic reaction A+B→C+D is given by:

$$k = <\sigma v> \quad (1)$$

where $\sigma$ is the total cross section of the reactants, $v$ is the relative velocity, and the averaging is performed over the thermal distribution. Most reactions have appreciable activation energy ($E_{\rm a}$) even if exothermic (this is because chemical reaction typically involve the breaking of chemical bonds before the formation of new ones), and the rate coefficients $k$ can be expressed by the simple Arrhenius formula (e.g. Herbst 1990):

$$k = A(T)\exp(-E_{\rm a}/k_{\rm B}T) \quad (2)$$

*Table 2.* Classes of gas–phase reactions

| Type | Process | Rate Coefficient |
|---|---|---|
| Ion–molecule | $A^+ + B \rightarrow C^+ + D$ | $\sim 10^{-9}$ cm$^3$ s$^{-1}$ |
| Dissociative Recombination | $AB^+ + e \rightarrow A + B$ | $\sim 10^{-6}$ cm$^3$ s$^{-1}$ |
| Radiative Association | $A + B \rightarrow AB + h\,\nu$ | $\sim 10^{-16}$–$10^{-9}$ cm$^3$ s$^{-1}$ |
| Neutral–neutral | $A + B \rightarrow C + D$ | $\sim 10^{-12}$–$10^{-10}$ cm$^3$ s$^{-1}$ |
| Photodissociation | $AB + h\nu \rightarrow A + B$ | $\sim 10^{-9}$ s$^{-1}$ |
| Charge–transfer | $A^+ + B \rightarrow A + B^+$ | $\sim 10^{-9}$ cm$^3$ s$^{-1}$ |

NOTE: Photodissociation rate listed for no extinction only.

where $k_B$ is the Boltzmann constant, $T$ is the gas kinetic temperature and $A(T)$ is the "pre–exponential factor", a weak function of the temperature, which depends upon the actual shape of the reaction potential surface. Activation energies of about 0.1–1 eV are normal, so that in dark clouds, where $k_BT \sim$ 0.01 eV, even exothermic reactions may not proceed.

Particularly important for the chemistry of dark clouds are ion–molecule reactions (Herbst & Klemperer 1973), which do not possess activation energy because of the strong long–range attraction force, given by the potential:

$$V(R) \quad = \quad -\frac{\alpha\, e^2}{2\, R^4} \tag{3}$$

where $\alpha$ is the polarizability of the neutral species (cm$^3$), $e$ is the electronic charge in esu, and $R$ is the distance between the two species. For a sufficiently close encounter, so that orbiting or spiraling occurs ($R \leq (4e^2\alpha/\mu v^2)^{1/4}$, where $\mu$ is the reduced mass in the collision), the rate coefficient becomes:

$$k_L \quad = \quad 2\pi e \left(\frac{\alpha}{\mu}\right)^{1/2} \quad \simeq \; 10^{-9}\,\mathrm{cm^3 s^{-1}}, \tag{4}$$

the so–called Langevin rate, independent of the temperature (Herbst 1990).

The main processes happening in the gas phase are listed in Tab. 2 along with their typical rate coefficients (see Tielens 2005 and Duley & Williams 1984 for details). Here it is worthwhile mentioning some examples, important for dark clouds and star forming regions.

*Ion–molecule reactions and dissociative recombination*

The production of ions in dark molecular clouds is initiated by cosmic rays ($c.r.$), which ionize $H_2$, the most abundant molecular species. The rate of this process, called $\zeta$, is typically of order $10^{-17}$ s$^{-1}$ (e.g. van der Tak & van Dishoeck 2000), and about 97% of the ionizations produce $H_2^+$:

$$H_2 + c.r. \quad \rightarrow \quad H_2^+ + e + c.r. \tag{5}$$

Once $H_2^+$ is formed, it reacts very quickly with $H_2$ to produce $H_3^+$, one of the most important molecular ions in astrochemistry:

$$H_2^+ + H_2 \rightarrow H_3^+ + H \tag{6}$$

In general, if an ion can react with $H_2$, no other reaction need to be considered because $H_2$ is at least three orders of magnitude more abundant than the other species and the reaction rate is proportional to the abundance of the reactants. $H_3^+$ does not react with $H_2$ but it can easily transfer a proton, thus increasing molecular complexity. For example, if $H_3^+$ encounters an atomic oxygen, $OH^+$ is produced ($H_3^+ + O \rightarrow OH^+ + H_2$). Then, successive "H atom" transfer (ion–$H_2$) reactions proceed until a saturated species (which cannot further react with $H_2$) is formed: $OH^+ \xrightarrow{H_2} H_2O^+ \xrightarrow{H_2} H_3O^+$. Upon *dissociative recombination* of the oxonium ion $H_3O^+$, water and OH form:

$$H_3O^+ + e \rightarrow H_2O + H \tag{7}$$

$$\rightarrow OH + 2H \text{ or } OH + H_2 \tag{8}$$

The fraction of $H_3O^+$ ions which will produce $H_2O$ and OH upon dissociative recombination is called the *branching ratio* and it is measured in the laboratory (e.g. Geppert et al. 2004). Ion–molecule chemistry of larger species leads to mainly H–poor (unsaturated) species, unlike dust chemistry (see Sec. 2).

*Neutral–neutral reactions*

Although they typically possess activation energy (and thus become important in high temperature regions; e.g. Sect. 2), some neutral–neutral reactions are especially crucial in dark clouds for nitrogen and sulphur chemistry, which starts with:

$$CH + N \rightarrow CN + H \tag{9}$$

$$S + OH \rightarrow SO + H \tag{10}$$

The reaction of CN with N will produce $N_2$ (also formed by NO + N), which can react with $H_3^+$ to form the easily observable $N_2H^+$. $N_2$ initiates the chemical path toward the production of another well known interstellar molecule: $NH_3$ ($N_2$ first reacts with $He^+$ to produce $N^+$, which is then successively hydrogenated by $H_2$ until the saturated $NH_4^+$ ion is formed; $NH_3$ is finally produced via $NH_4^+$ dissociative recombination). The common chemical origin between $N_2H^+$ and $NH_3$ explains the good correlation between the two species found in molecular cloud cores (Benson et al. 1998).

## Surface chemistry

The fundamental role played by dust grains in synthesizing molecular species is well known for about fourty years, when they were invoked to explain the production rate of $H_2$ molecules in our Galaxy. Pioneer work in this field has been done by Gould & Salpeter (1963) and Hollenbach & Salpeter (1971), who first derived the formation rate of $H_2$ on grain surfaces. From observations carried out by the Copernicus satellite, the derived rate of formation for $H_2$ in diffuse clouds is $R \simeq 3\times10^{-17}$ cm$^3$s$^{-1}$ (Jura 1975). Lots of theoretical and laboratory work has been done since then on this fundamental process (e.g. Pirronello et al. 1999; Williams et al. 2000; Cazaux & Tielens 2004). Observations of the so–called HINSA (HI Narrow Self–Absorption) in cloud cores are also giving clues on the $H_2$ formation efficiency in regions where dust grains are covered by icy mantles (e.g. Li & Goldsmith 2005).

Surface chemistry is also invoked to explain ($i$) the observed large abundances of complex species near star forming regions (Sect. 2); ($ii$) the orders–of–magnitude abundance enhancements of selected molecular species observed along molecular outflows (Sect. 2); ($iii$) dust grain mantles composed of complex icy mixtures, as deduced from ground–based and space – in particular, ISO, the Infrared Space Observatory – observations of background field stars or embedded protostars (e.g. Sellgren et al. 1994; Gibb et al. 2004; van Dishoeck 2004); ($iv$) the large amount of deuterium fractionation observed in low mass star forming regions (Sect. 1 and 2).

The main surface processes are illustrated in Fig. 1 (from Herbst 2000), which shows a portion of an idealized grain surface, with adjacent sites. Once a light atom (such as H) hits the dust surface, it is first *adsorbed* in a binding site. The species is *physisorbed* if it is linked to the surface by van der Waals forces, and binding energies are in the range 0.01–0.2 eV (100–2000 K). In this case, light species can quickly ($<<$ 1 s) move across the surface, overcoming the energetic barriers via thermal hopping or quantum tunneling, depending upon the dust temperature (e.g. Tielens & Allamandola 1987). Indeed, surface chemistry is not limited by reaction rates but by the *accretion rates* of reactive species on grain surfaces (e.g. in a typical dark cloud, one H atom hits one dust grain approximately once every 10 days).

Surface chemistry based on diffusion is known as a *Langmuir–Hinshelwood* (LH) chemistry (see also Tielens & Hagen 1982). Here, *hydrogenation* is one of the main surface processes, so that O is transformed to $H_2O$ (the main ice component), C to $CH_4$, N to $NH_3$, S to $H_2S$, and CO to $H_2CO$ and $CH_3OH$. Hydrogenation of CO implies the overcoming of an activation energy barrier (Woon 2002), which slows down the process; in fact, CO is one of the major compounds of grain mantles observed in molecular clouds, second only to water (e.g. Chiar et al. 1995; Teixeira et al. 1998). There are still some doubts about the existence of solid $NH_3$ (e.g. Schutte et al. 2003) and $H_2S$ (see discussion in Wakelam et al. 2004). But solid OCS has been detected (Palumbo et al. 1997) and $CO_2$ absorption features are ubiquitous in the direction of background and embedded objects (e.g. Nummelin et al. 2001). Therefore,

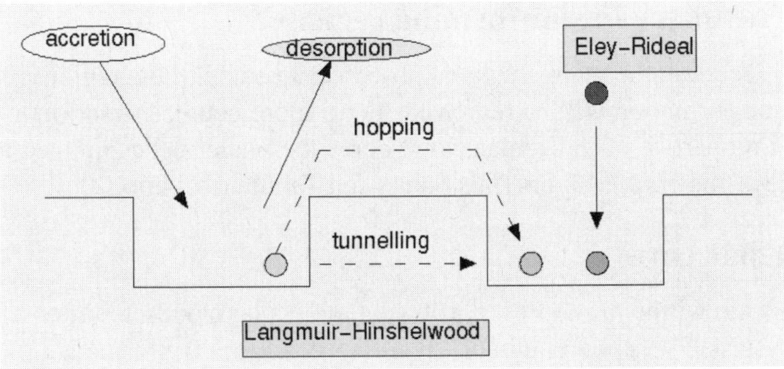


*Figure 1.* Assorted processes on dust grain surfaces (from Herbst 2000).

oxygenation is likely to be another important surface process, which competes with hydrogenation depending on the gaseous O/H ratio and the dust temperature (H being more volatile than O).

If the accreted species are strongly bound, or *chemisorbed*, with typical strengths of $\sim$ 1 eV (10000 K), the *Eley–Rideal* (ER) mechanism dominates (see Fig. 1): a surface reaction only occurs when a gas–phase species lands atop or nearby the chemisorbed species. This type of chemistry, efficient only for high surface coverage, preferentially occurs at dust temperatures $T_{\rm d}$ below about 300 K (for a barrier between chemisorption sites of $\sim$1 eV, S. Cazaux, priv. comm.), when the LH mechanism cannot proceed for chemisorbed species (see discussion in Habart et al. 2004 on the importance of the various mechanisms for $H_2$ formation in photodissociation regions).

The return of surface species back in the gas phase can be driven by *thermal desorption*, with a time scale exponentially dependent on the dust temperature. At $T_{\rm d}$ = 10 K, only H and D can thermally evaporate with significantly large rates, whereas $T_{\rm d} > 90$ K is needed to start $H_2O$ evaporation. Thus, large abundances of water and other saturated species are expected in the gas phase in star forming regions where the dust has been heated above, say, 100 K by the protostellar radiation field (see Sect. 2). Among the main *nonthermal desorption* mechanisms, important in cold clouds, are: ($i$) cosmic–ray bombardment (especially by the heavier nuclei; Léger et al. 1985; Hasegawa & Herbst 1993), and ($ii$) the energy generated by chemical surface reactions, still debatable (Willacy & Millar 1998; Takahashi & Williams 2000). UV photons play a role in photodesorption of solid species in PDRs and diffuse clouds (although the yield of photodesorption is still not well constrained; e.g. Ruffle & Herbst 2001).

## Astrochemistry in star forming regions

In this section, the main chemical processes in some of the regions of interest for star formation will be reviewed. For a more detailed description of the physical properties of molecular cloud cores the reader is encouraged to read Walmsley (this book), Myers (this book) and Tafalla (this book).

## Pre–stellar cores

Stars form within the densest portions of molecular clouds, the "dense cores" (e.g. Myers 1999), upon gravitational collapse. Before the formation of a protostellar object, (starless) dense cores are characterized by low temperatures ($\sim$ 10 K) and relatively high central densities ($\simeq 10^5$ cm$^{-3}$). These conditions favour the freeze–out of gas phase species onto dust grains. This can be easily understood, comparing the freeze out and the dynamical time scales. The time for a gaseous species X to accrete onto a dust grain is given by:

$$t_{\rm fo} = \frac{1}{S n_{\rm d} \pi a_{\rm d}^2 v_{\rm t}} \quad (11)$$

$$\simeq 10^9 \sqrt{m_{\rm X}/T} (n_{\rm H} S)^{-1} \ {\rm yr}, \quad (12)$$

where S is the sticking coefficient or the probability of a certain species to stick on the grain upon collision (about unity), $n_{\rm d}$ is the number density of dust grains, $a_{\rm d}$ is the grain radius, $v_{\rm t}$ is the thermal velocity of the gaseous species, $n_{\rm H}$ is the number density of hydrogen nuclei (it has been assumed that the gas–to–dust density ratio (by mass) is 100 and $a_{\rm d}$ = 0.1 $\mu$m), and $m_{\rm X}$ is the mass of species X in amu. Therefore, in a cloud with number density $n_{\rm H}$ = $10^5$ cm$^{-3}$, $t_{\rm fo} \sim 10^4$ yr. The free–fall time scale, the shortest dynamical time scale, for a cloud with the same $n_{\rm H}$ value is (e.g. Spitzer 1978):

$$t_{\rm ff} = 4 \times 10^7 (n_{\rm H})^{-1/2} \ ({\rm yr}), \quad (13)$$

or $10^5$ yr. Thus, freeze out is expected to be an important process in the evolution of cloud cores toward the formation of a star.

In fact, strong depletion of CO from the gas phase has been observed in a variety of molecular cloud cores (e.g. Caselli et al. 1999; Bergin et al. 2001; Tafalla et al. 2004). Tafalla (this volume) presents a detailed review of the observed chemical properties of starless cores. Here, I summarize the main observed properties of *pre–stellar cores*, which differ from starless cores in being more centrally concentrated and showing kinematic evidences of central infall (Ward–Thompson et al. 1999; Crapsi et al. 2005). Pre–stellar cores are thought to be unstable objects, on the verge of star formation, thus their study is crucial to unveil the initial conditions of the star formation process. Fig. 2 shows a schematic picture of the chemical structure of a "typical" pre–stellar

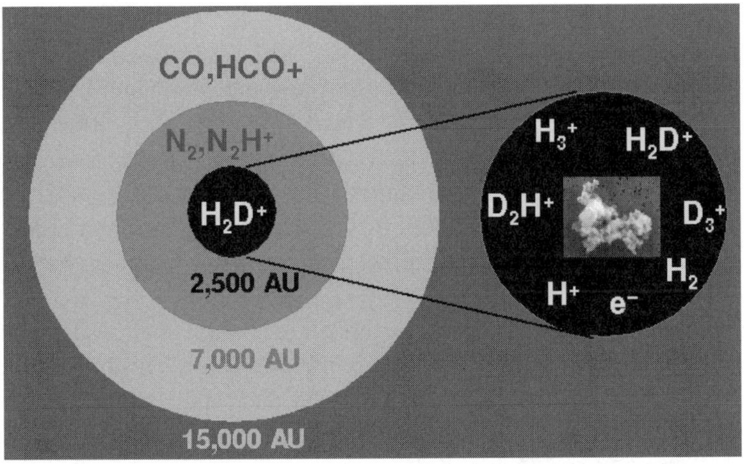


*Figure 2.* Schematic picture of the chemical structure of a typical pre–stellar core. The central density is around $10^6$ cm$^{-3}$ and it drops to $\sim 2\times10^4$ cm$^{-3}$ at the edge. In the outer shell, CO is mostly in the gas phase and $HCO^+$ is the main ion. Within about 7,000 AU, CO is mainly frozen onto dust grains whereas $N_2$ remains in the gas phase, so that a good probe of these zones is $N_2H^+$. In the central $\sim$ 2,500 AU (the "molecular hole"), only light species survive in the gas phase. Here, the thick icy mantle may boost the coagulation of dust grains.

core: $(i)$ the core envelope ($\sim$ 7,000–15,000 AU): CO is mainly in the gas phase, and the main molecular ion is $HCO^+$; $(ii)$ the core nucleus ($\sim$ 2,500–7,000 AU): CO and other C–bearing species are highly depleted from the gas phase, unlike N–bearing species (e.g. Bergin & Langer 1987); thus, $N_2H^+$ is a good tracer of these zones and the deuterium fractionation is very large (see below); $(iii)$ the "molecular hole" (within $\sim$ 2,500 AU): all species heavier than helium are likely to be condensed out onto dust grains (the best tracer of this zone may be $H_2D^+$).

*Freeze–out and deuterium fractionation*

An important consequence of the depletion of gas phase species is the large enhancement in the deuterium fractionation of molecular species, which has been extensively observed in the past years (e.g. Caselli et al. 2002; van der Tak et al. 2002; Bacmann et al. 2003). This process involves the deuteron–proton exchange reaction (Watson 1974):

$$H_3^+ + HD \rightarrow H_2D^+ + H_2 + \Delta E, \quad (14)$$

with $\Delta E \simeq 230$ K, which can only proceed from left to right in cold dark clouds. Here, in fact, the $H_2D^+/H_3^+$ abundance ratio increases well above the cosmic D/H ratio ($\sim 1.5\times10^{-5}$). $H_2D^+$ can react with CO, $N_2$ and other neutrals, transferring the deuteron in at least one third of its encounters, with the consequence of enhancing the deuterium fractionation in, e.g., $HCO^+$ and $N_2H^+$, as observed (e.g. Williams et al. 1998). In the case of CO (and other

$H_3^+$ and $H_2D^+$ destruction partners) depletion, the $H_3^+$ abundance will increase and, consequently, the $H_2D^+$ production rate will raise (via reaction 14), further increasing the $H_2D^+/H_3^+$ abundance ratio and the deuterium fractionation. In pre–stellar cores, $N_2D^+/N_2H^+ \simeq 0.2$ (Caselli et al. 2002; Crapsi et al. 2005) and $D_2CO/H_2CO \simeq 0.05$ (Bacmann et al. 2003), orders of magnitude larger that the cosmic D/H ratio. The observed D–fractionation also correlates with the measured amount of CO depletion (Bacmann et al. 2003; Crapsi et al. 2005), as expected from theory (e.g. Dalgarno & Lepp 1984; Aikawa et al. 2005).

In *molecular holes*, volume densities are large enough to allow an "extreme" depletion of gaseous species heavier than helium. This seems to be supported by the detection of a strong $H_2D^+$ line in the direction of L1544 (Caselli et al. 2003). The estimated $H_2D^+$ abundance cannot be reproduced by standard gas–grain chemistry unless CO is mainly in solid form (e.g. Roberts et al. 2003; Walmsley et al. 2004). If this is the case, the only tracer of the chemical composition of this region may be $H_2D^+$. Also $D_2H^+$ has been observed in another pre–stellar core (Vastel et al. 2004), with abundances similar to $H_2D^+$, which further pointed out the importance of including multiply deuterated species in chemical models of pre–stellar cores.

## Outflows

In their youngest phase, protostars are called Class 0 sources (André et al. 1993) and one of their main characteristics is the presence of powerful and collimated outflows (e.g. Bachiller 1996; see also Beuther & Shepherd, this book and Bacciotti, this book). Outflows shock the material along their path, raising the gas temperature and partially destroying dust grains. In the presence of magnetic fields, if the electron fraction is sufficiently low (as in the dense material surrounding Class 0 sources), and if the shock velocity is below about 50 km s$^{-1}$ (Draine et al. 1983), a discontinuity in the hydrodynamical variables (e.g. density and temperature) is not present and the shock is named "C–type" ("C" stays for "continuous"). In all the other cases, a "jump front" is formed, and the shock is called "J–type" (see Draine 1980).

In "C–type" shocks, molecular species (in particular $H_2$) do not dissociate. In these conditions, and if the gas temperature raises above $\sim$200 K, some important endothermic reactions become fast and quickly convert most of the free oxygen into water ($O \xrightarrow{H_2} OH \xrightarrow{H_2} H_2O$), and sulphur mainly into SO and $SO_2$ (e.g. Hartquist et al. 1980; Pineau des Forêts et al. 1993). In dissociative shocks, as soon as the post–shocked gas temperature has fallen to a few hundred degrees, $H_2$ molecules are reformed and shock chemistry proceeds as in C–type shocks (e.g. Neufeld & Dalgarno 1989).

The presence of magnetic fields in C–type shocks implies different velocities for neutrals and charged particles, including dust grains which are negatively charged and linked to the magnetic field lines. In these conditions, the collisions between (especially) $H_2$ molecules and dust grains may cause the return of mantle and refractory species in the gas phase. This process is called grain *sputtering* (e.g. Flower & Pineau des Forêts 1995; Schilke et al. 1997) and it is thought to be responsible for the observed enhancement of SiO and saturated molecular species such as $H_2O$, $NH_3$, and $CH_3OH$ along protostellar outflows (e.g. Bachiller & Perez Gutierrez 1997; Jørgensen et al. 2004). Dust grains with different sizes will also have relative velocities because of the different mass and charge (entering the gyrofrequency expression), so that grain–grain collisions are also quite efficient in vaporising mantle and refractory material (Caselli et al. 1997).

Once icy mantles and part of the refractory material are released in the gas phase, shock chemistry proceeds until the gas temperature drops to the pre–shock values. This happens in a short time scale (a few hundred years), because of the efficient gas cooling mainly due to $H_2$, water and CO (e.g. Kaufman & Neufeld 1996). Then, the post–shocked gas will experience again the "cold chemistry", including freeze–out (e.g. Bergin et al. 1998). This mechanism (shock+freeze–out) has been suggested by Bergin et al. (1999) as a way to explain the large abundance ($\sim 10^{-4}$ w.r.t. $H_2$) of water in solid form in molecular clouds.

## Hot Cores

The newborn protostar eventually heats up the surrounding material with its radiation, raising the dust temperature above the sublimation value for grain mantles, and forming the so–called "Hot Cores" (HCs). HCs are typically associated with high mass star forming regions and represent a stage earlier than ultracompact HII regions. They have diameters of $<$0.1 pc, $H_2$ number densities of $\geq 10^7$ cm$^{-3}$ and gas temperatures $\geq$100 K (e.g. Kurtz et al. 2000). Their chemical signatures are: $(i)$ large abundances of saturated species, such as $H_2O$, $NH_3$, $H_2S$, $CH_3OH$ (e.g. Pauls et al. 1983; Menten et al. 1986), suggestive of an active surface chemistry (Sect. 2); $(ii)$ large abundances of complex organic H–rich molecules, in particular $HCOOCH_3$ (methyl formate), $CH_3COOH$ (acetic acid), $CH_3CH_2CN$ (ethyl cyanide), $CH_3CH_2OH$ (ethanol), $(CH_3)_2CO$ (acetone) (e.g. Millar et al. 1995; Mehringer & Snyder 1996; Remijan et al. 2004), hard to form in the gas phase; $(iii)$ relatively large deuterium fractionations (e.g. Oloffson 1984; Turner 1990), a record of an early cold phase (Sect. 1). Interestingly, these regions show a strong chemical differentiation, especially evident in the nearby Orion KL (e.g. Blake et al. 1987;

Wright et al. 1996) and also in the W3(OH) region (Wyrowski et al. 1997), when millimeter interferometry is used.

To understand the gas phase composition in HCs, surface chemistry is needed, because gas phase routes appear to be too slow. In particular, saturated species can be efficiently formed on grain surfaces before or during the earliest stages of protostar evolution, when the dust temperature is low enough ($\leq$ 20K) to allow hydrogenation and produce solid $H_2O$, $NH_3$, $CH_3OH$ and $H_2S$ (see Sect. 2). Once the dust is heated to temperatures larger than the sublimation values for icy mantles (about 100 K), these saturated species will enrich the gas phase and start a hot gas phase chemistry (e.g. Brown et al. 1988; Charnley et al. 1992). A way to reproduce the observed chemical differentiation in the Orion KL region, where N–bearing and O–bearing species are spatially segregated, is to take into account the temperature and density gradients of the circumstellar cloud (Caselli et al. 1993).

In general, different routes are invoked for complex O–bearing and N–bearing species: the former are produced after the evaporation of solid methanol ($CH_3OH$) and formaldehyde ($H_2CO$) (the complexity is increasing in the gas phase); the latter are formed on grain surfaces (e.g. ethyl cyanide, $CH_3CH_2CN$) and, once evaporated, they form simpler daughter species (e.g. vinyl cyanide, $CH_2CHCN$). However, as recently found in the laboratory (Horn et al. 2005), some key reactions in the chain to form the HC–ubiquitous methyl formate from evaporated methanol do not proceed. Thus, surface chemistry is also needed to produce complex O–bearing species. *This is all to be explored!* In fact, current chemical models neglect surface chemistry, although they are very sophisticated in treating the physical structure of the molecular clouds (e.g. Doty et al. 2002; Rodgers & Charnley 2003; Nomura & Millar 2004; Viti et al. 2004).

In the past few years it has been realized that HCs are not only a signature of high mass star formation, but they are also found in low mass star forming regions, during the earliest (Class 0) protostellar phase (Cazaux et al. 2003; Bottinelli et al. 2004a,b; Kuan et al. 2004), and in one intermediate mass Class 0 source (Fuente et al. 2005). Low–mass HCs, called "hot corinos", are a "scaled" version ($n(\mathrm{H_2}) \sim 10^7$ cm$^{-3}$, $T \sim$ 50–100 K, size $\sim$ 150 AU) of the massive HCs. Nevertheless, they show rich spectra at millimeter wavelengths, with complex organic molecules, which is challenging current theory (see e.g. Rodgers & Charnely 2001). Outside these "hot corinos", the gas and the dust still maintain the pre–stellar–core characteristics, including freeze–out (e.g. Belloche & André 2004). In the well–studied case of IRAS 16293-2422, the deuterium fractionation is very large (especially for $H_2CO$ and $CH_3OH$; e.g. Parise et al. 2004) compared to high mass HCs, probably suggesting a different chemical evolution, such as lower dust temperatures during the pre–stellar phase, which favour surface deuteration, or different chemical time scales.

## Protoplanetary Disks

Protoplanetary disks around T Tauri and Herbig Ae/Be stars, which are contracting toward the main sequence and have ages of a few millions years, are an important link between molecular clouds and protoplanetary systems (see Markwick & Charnley 2004 for a recent review). Understanding their chemical composition and the chemical processes is in fact one of the major challenge for astrochemistry, although the next generation of telescopes (in particular ALMA, the Atacama Large Millimeter Array) is needed to put stringent constraints on the models. So far, observations of species such as CO, CS, CN, HCN, $HCO^+$, $N_2H^+$, and $H_2CO$ have shown that molecular abundances are typically lower than those in dense interstellar clouds, suggesting that molecular freeze–out is at work (e.g. Dutrey et al. 1997; Goldsmith et al. 1999; Qi et al. 2003; Thi et al. 2004). Moreover, $DCO^+$ has also been detected and the derived deuterium fractionation is comparable to that observed in dense cloud cores ($DCO^+/HCO^+ \sim 0.04$; van Dishoeck et al. 2003). Ceccarelli et al. (2004) have detected $H_2D^+$, known to be abundant in highly depleted regions (see Sect. 1). There is also observational evidence that photon–dominated chemistry is significant (because of the relatively large abundance of CN and $C_2H$, e.g. Dutrey et al. 1997). Theoretical work has pointed out the importance of X–rays in the ionization (e.g. Glassgold et al. 1997) and thermal (e.g. Glassgold et al. 2004) balance. Thus, many ingredients participate in the chemistry of protoplanetary disks.

In current chemical models, protoplanetary disks can be divided in three layers (see Fig. 3): the *surface layer*, where photochemical processes are important because of the exposure to the UV flux from the star and the interstellar radiation field; the *intermediate layer*, where the extinction is large enough that the chemistry resembles that of dense clouds; the *midplane*, where the density is so large that the majority of gas phase species are frozen onto dust grains (see, e.g., Aikawa et al. 2001; Willacy & Langer 2000; Markwick et al. 2002; van Zadelhoff et al. 2003; Semenov et al. 2004).

## Conclusions

Tab. 3 summarizes the main processes for the chemistry of star forming regions, in the three phases (pre–stellar cores, Class 0 protostars and protoplanetary disks) briefly illustrated in this review. Question marks in the table follow not–yet–proved statements, mostly based on model predictions.

Here, I would like to point out some of the main problems still open in the field of astrochemistry of star forming regions. Many uncertainties are present especially regarding surface chemistry (e.g. Stantcheva & Herbst 2004; Chang et al. 2004) and gas phase reactions at high temperature (e.g. Wakelam et al. 2004), which in particular affect our interpretation of the observations to-

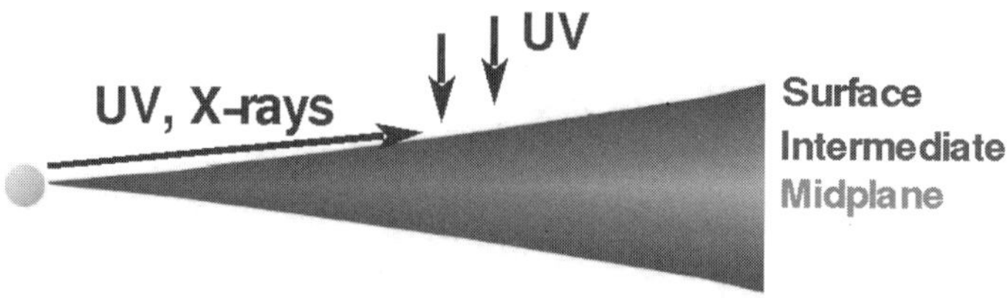


*Figure 3.* Schematic picture of a protoplanetary disk, where three chemically distinct zones can be recognized: the *midplane*, where freeze–out is dominant; the *intermediate zone*, where dense cloud chemistry is appropriate; the *surface*, where photochemistry is important (courtesy of Y. Aikawa).

*Table 3.* Main processes and molecular gas tracers in star forming regions

| Region | Gas Tracer | Main Processes |
|---|---|---|
| *Pre–stellar cores*: | | |
| envelope | CO, $HCO^+$, CS | ion–molecule |
| nucleus | $N_2H^+$, $N_2D^+$, $NH_3$ | freeze–out, D–fractionation |
| molecular hole | $H_2D^+$, $D_2H^+$, $D_3^+$? | freeze–out, grain coagulation? |
| *Class 0*: | | |
| envelope | $N_2H^+$, $NH_3$ | ion–molecule, freeze–out |
| outflow | $H_2O$, SiO, $CH_3OH$, $SO_2$ | grain sputtering, neutral–neutral |
| hot cores | $H_2O$, complex organics | mantle evaporation, neutral–neutral |
| *Protoplanetary disks*: | | |
| surface layer | CO, CN, $C_2H$ | photochemistry, ion–molecule |
| intermediate layer | HCN, $DCO^+$, $N_2H^+$ | freeze–out, D–fractionation |
| midplane | $H_2D^+$, $H^+$?, $D_2H^+$?, $D_3^+$? | freeze–out, grain coagulation |

ward molecular outflows and hot cores. Some important problems to solve are: $(i)$ after recent laboratory work (Horn et al. 2004; Luca et al. 2004), we are left with the non–trivial problem of producing $HCOOCH_3$ and $CH_3OH$ in the gas phase. It seems that the formation on grain surfaces (not yet explored for methyl formate) followed by desorption is the only possible route. $(ii)$ Grain mantle composition is expected to change during cloud evolution and contraction; thus, binding energies (and consequently freeze–out rates) are a function of time. Nevertheless, current models simply neglect this. $(iii)$ There are still quite large uncertainties regarding the chemistry of Oxygen (e.g. how much atomic O is left in the gas phase in pre–stellar core?), and Sulphur (where does S go in dense clouds and what is its form in icy grain mantles?). $(iv)$ How good is the approximation of assuming a constant cosmic ray ionization rate ($\zeta$) throughout the core? A large value of $\zeta$ ($\sim 10^{-15}$ s$^{-1}$) has been deduced for diffuse clouds (McCall et al. 2003). But this value drops to 1–3$\times 10^{-17}$ s$^{-1}$ in dense clouds; does it continue to drop *within* pre–stellar cores, with a consequent variation of the fractional ionization and the chemistry in cloud core nuclei? A synergy of laboratory, theoretical, and observational work, as well as the new generation of telescopes and interferometers (ALMA, SMA, CARMA, Herschel, SOFIA, APEX), are sorely needed to advance our understanding of astrochemistry.

**Acknowledgements**

I thank all my collaborators, who keep feeding my enthusiasm about astrochemistry. A special thank to Eric Herbst and Malcolm Walmsley to have had the patience of going through this review and giving me important input. Finally, I thank all the organizers of this meeting and, in particular, Nanda Kumar for all his hard work, needed to make the *C2C* a very successful (and enjoyable) meeting.

## References

Aikawa, Y., Herbst, E. (2001). A&A, 371, 1107

Aikawa, Y., Herbst, E., Roberts, H., Caselli, P. (2005). ApJ, 620, 330

Anders, E., Grevesse, N. (1989). Geochimica et Cosmochimica Acta, 53, 197

Andrè, P., Ward-Thompson, D. Barsony, M. (1993). ApJ, 406, 122

Asplund, M., Grevesse, N., Sauval, J. (2005). In: *Cosmic abundances as records of stellar evolution and nucleosynthesis* ed by F.N. Bash & T.G Barnes (ASP conf. series), in press (astro-ph/0410214)

Bachiller, R. (1996). ARA&A, 34, 111

Bachiller, R., Perez Gutierrez, M. (1997). ApJ, 487, L93

Bacmann, A., Andrè, P., Puget, J.-L., Abergel, A., Bontemps, S., Ward-Thompson, D. (2000). A&A, 361, 555

Bacmann, A., Lefloch, B., Ceccarelli, C., Steinacker, J., Castets, A., Loinard, L. (2003). ApJ, 585, L55

Bell, M. B., Matthews, H. E. (1985). ApJ, 291, L63

Belloche, A., André, P. (2004). A&A, 419, L35
Benson, P. J., Caselli, P., Myers, P. C. (1998). ApJ, 506, 743
Bergin, E. A., Ciardi, D. R., Lada, C. J., Alves, J., Lada, E. A. (2001). ApJ, 557, 209
Bergin, E. A., Langer, W. D. (1997). ApJ, 486, 316
Bergin, E. A., Neufeld, D. A., Melnick, G. J. (1998). ApJ, 499, 777
Bergin, E. A., Neufeld, D. A., Melnick, G. J. (1999). ApJ, 510, L145
Bianchi, S., Gonąlves, J., Albrecht, M., Caselli, P., Chini, R., Galli, D., Walmsley, C. M. (2003). A&A, 399, L43
Blake, G.A., Sutton, E.C., Masson, C.R., Phillips, T.G. (1987). ApJ, 315, 621
Bottinelli, S., Ceccarelli, C., Lefloch, B., et al. (2004). ApJ, 615, 354
Bottinelli, S., Ceccarelli, C., Neri, R., et al. (2004). ApJ, 617, L69
Brown, P. D., Charnley, S. B., Millar, T. J. (1988). MNRAS, 231, 409
Caselli, P., Hartquist, T. W., Havnes, O. (1997). A&A, 322, 296
Caselli, P., Hasegawa, T. I., Herbst, Eric (1993). ApJ, 408, 548
Caselli, P., van der Tak, F. F. S., Ceccarelli, C., Bacmann, A. (2003). A&A, 403, L37
Caselli, P., Walmsley, C. M., Tafalla, M., Dore, L., Myers, P. C. (1999). ApJ, 523, L165
Caselli, P., Walmsley, C. M., Zucconi, A., Tafalla, M., Dore, L., Myers, P. C. (2002). ApJ, 565, 344
Ceccarelli,C., Dominik,C., Lefloch,B., Caselli,P., Caux, E. (2004). ApJ,607,L51
Cazaux, S., Tielens, A. G. G. M. (2004). ApJ, 604, 222
Cazaux, S., Tielens, A. G. G. M., Ceccarelli, C., Castets, A., Wakelam, V., Caux, E., Parise, B., Teyssier, D. (2003). ApJ, 593, L51
Chang, Q., Cuppen, H. M., Herbst, E. (2005). A&A, in press
Charnley, S. B., Tielens, A. G. G. M., Millar, T. J. (1992). ApJ, 399, L71
Chiar, J.E., Adamson, A.J., Kerr, T.H., Whittet, D.C.B. (1995). ApJ, 455, 234
Crapsi, A., Caselli, P., Walmsley, C. M., Myers, P. C., Tafalla, M., Lee, C. W., Bourke, T. L. (2005). ApJ, 619, 379
Dalgarno, A., Lepp, S. (1984). ApJ, 287, L47
Doty, S. D., van Dishoeck, E. F., van der Tak, F. F. S., Boonman, A. M. S. (2002). A&A, 389, 446
Draine, B. T. (1980). ApJ, 241, 1021
Draine, B. T., Roberge, W. G., Dalgarno, A. (1983). ApJ, 264, 485
Duley, W. W., Williams, D. A. (1984). *Interstellar chemistry*, London, England and Orlando, FL, Academic Press
Dutrey, A., Guilloteau, S., Guelin, M. (1997). A&A, 317, L55
Ehrenfreund, P., Charnley, S. B. 2000, ARA&A, 38, 427
Evans, N. J., II (1999). ARA&A, 37, 311
Evans, N. J., II, Rawlings, J. M. C., Shirley, Y. L., Mundy, L. G. (2001). ApJ, 557, 193
Flower, D. R., Pineau des Forêts, G. (1995). MNRAS, 275, 1049
Fuente, A., Rizzo, J. R., Caselli, P., Bachiller, R., Henkel, C. (2005). A&A, in press (astro–ph/0411602)
Galli, D., Walmsley, C. M., Gonąlves, J. (2002). A&A, 394, 275
Geppert, W. D., Thomas, R., Semaniak, J., et al. (2004). ApJ, 609, 459
Gibb, E. L., Whittet, D. C. B., Boogert, A. C. A., Tielens, A. G. G. M. (2004). ApJ, 610, L113
Glassgold, A. E., Najita, J., Igea, J. (1997). ApJ, 480, 344
Glassgold, A. E., Najita, J., Igea, J. (2004). ApJ, 615, 972
Goldsmith, P. F., Langer, W. D., Velusamy, T. (1999). ApJ, 519, L173
Gould, R. J., Salpeter, E. E. (1963). ApJ, 138, 393
Graedel, T. E., Langer, W. D., Frerking, M. A. (1982). ApJS, 48, 321

Habart, E., Boulanger, F., Verstraete, L., Walmsley, C. M., Pineau des Forêts, G. (2004). A&A, 414, 531

Hartquist, T. W., Dalgarno, A., Oppenheimer, M. (1980). ApJ, 236, 182

Hartquist, T. W., Williams, D. A. (1995). *The Chemically Controlled Cosmos, Astronomical Stars* (Cambridge University Press)

Hartmann, L. (2003). ApJ, 585, 398

Hasegawa, T. I., Herbst, E. (1993). MNRAS, 261, 83

Hasegawa, T. I., Herbst, E., Leung, C. M. (1992). ApJS, 82, 167

Herbst, E. (1990). Angew. Chem. Int. Ed. Engl., 29, 595

Herbst, E. (2000). In: *Astrochemistry: From Molecular Clouds to Planetary Systems*, ed by Y. C. Minh & E. F. van Dishoeck, p.147

Herbst, E., Klemperer, W. (1973). ApJ, 185, 505

Herbst, E., Leung, C. M. (1989). ApJS, 69, 271

Hollenbach, D., Salpeter, E. E. (1971). ApJ, 163, 155

Horn, A., Møllendal, H., Sekiguchi, O., Uggerud, E., Roberts, H., Herbst, E., Viggiano, A. A., Fridgen, T. D. (2004). ApJ, 611, 605

Jones, A. P., Williams, D. A. (1985). MNRAS, 217, 413

Jørgensen, J. K., Hogerheijde, M. R., Blake, G. A., van Dishoeck, E. F., Mundy, L. G., Schïer, F. (2004). A&A, 415, 1021

Jura, M. (1975). ApJ, 197, 575

Kaufman, M. J., Neufeld, D. A. (1996). ApJ, 456, 250

Kuan, Y.-J., Huang, H.-C., Charnley, S. B., et al. (2004). ApJ, 616, L27

Kurtz, S., Cesaroni, R., Churchwell, E., Hofner, P., Walmsley, C. M. (2000). In: *Protostars and Planets IV*, eds Mannings, V., Boss, A.P., Russell, S. S., p. 299 (University of Arizona Press)

Lee, C. W., Myers, P. C., Tafalla, M. (2001). ApJS, 136, 703

Léger, A., Jura, M., Omont, A. (1985). A&A, 144, 147

Leung, C. M., Herbst, E., Huebner, W. F. (1984). ApJS, 56, 231

Li, D., Goldsmith, P. F. (2005). ApJ, in press (astro-ph/0412427)

Luca, A., Voulot, D., Gerlich, D. *WDS'02 Proceedings of Contributed Papers, Part II*, 2002, MATFYZPRESS, p. 294

Markwick, A. J., Charnley, S. B. (2004). In: *Astrobiology: Future Perspectives*, by P.Ehrenfreund et al. Leiden Observatory, The Netherlands Astrophysics and Space Science Library, Vol. 305 (Kluwer Academic Publishers), p. 33

Markwick, A.J., Ilgner, M., Millar, T.J., Henning, Th. (2002). A&A, 385, 632

McCall, B. J., Huneycutt, A. J., Saykally, R. J., et al. (2003). Nature, 422, 500

McKee, C. F. (1989). ApJ, 345, 782

Menten, K. M., Walmsley, C. M., Henkel, C., Wilson, T. L., Snyder, L. E., Hollis, J. M., Lovas, F. J. (1986). A&A, 169, 271

Mehringer, D. M., Snyder, L. E. (1996). ApJ, 471, 897

Millar, T. J., MacDonald, G. H., Habing, R. J. (1995). MNRAS, 273, 25

Morton, D. C. (1975). ApJ, 197, 85

Myers, P. C. (1999). In: *The Origin of Stars and Planetary Systems*, ed by C. J. Lada and N. D. Kylafis, p.67 (Kluwer Academic Publishers)

Neufeld, D. A., Dalgarno, A. (1989). ApJ, 340, 869

Nomura, H., Millar, T. J. (2004). A&A, 414, 409

Nummelin, A., Whittet, D. C. B., Gibb, E. L., Gerakines, P. A., Chiar, J. E. (2001). ApJ, 558, 185

Ohashi, N., Lee, S. W., Wilner, D. J., Hayashi, M. (1999). ApJ, 518, L41

Olofsson, H. (1984). A&A, 134, 36

Ossenkopf, V., Henning, Th. (1994). A&A, 291, 943

Pagani, L., Bacmann, A., Motte, F., et al. (2004). A&A, 417, 605
Palla, F., Stahler, S. W. (2002). ApJ, 581, 1194
Palumbo, M. E., Geballe, T. R., Tielens, A. G. G. M. (1997). ApJ, 479, 839
Parise, B., Castets, A., Herbst, E., Caux, E., Ceccarelli, C., Mukhopadhyay, I., Tielens, A. G. G. M. (2004). A&A, 416, 159
Pauls, A., Wilson, T. L., Bieging, J. H., Martin, R. N. (1983). A&A, 124, 23
Pineau des Forêts, G., Roueff, E., Schilke, P., Flower, D. R. (1993). MNRAS, 262, 915
Pirronello, V., Liu, C., Roser, J. E., Vidali, G. (1999). A&A, 344, 681
Prasad, S. S., Huntress, W. T., Jr. (1982). ApJ, 260, 590
Qi, C., Kessler, J. E., Koerner, D. W., Sargent, A. I., Blake, G. A. (2003). ApJ, 597, 986
Remijan, A., Shiao, Y.-S., Friedel, D. N., Meier, D. S., Snyder, L. E (2004). ApJ, 617, 384
Roberts, H., Herbst, E., Millar, T. J. (2003). ApJ, 591, L41
Rodgers, S. D., Charnley, S. B. (2001). ApJ, 546, 324
Rodgers, S. D., Charnley, S. B. (2003). ApJ, 585, 355
Ruffle, D. P., Hartquist, T. W., Caselli, P., Williams, D. A. (1999). MNRAS, 306, 691
Ruffle, D. P., Herbst, E. (2001). MNRAS, 322, 770
Schilke, P., Walmsley, C. M., Pineau des Forêts, G., Flower, D. R. (1997). A&A, 321, 293
Schutte, W. A., Khanna, R. K. (2003). A&A, 398, 1049
Sellgren, K., Smith, R. G., Brooke, T. Y. (1994). ApJ, 433, 179
Semenov, D., Wiebe, D., Henning, Th. (2004). A&A, 417, 93
Shinnaga, H., Ohashi, N., Lee, S.-W., Moriarty-Schieven, G. H. (2004). ApJ, 601, 962
Snow, T. P., Witt, A. N. (1996). ApJ, 468, L65
Spitzer, L. (1978). *Physical processes in the interstellar medium* (New York Wiley-Interscience)
Stahler, S. W., Palla, F. 2004, *The formation of stars*, New York, NY: Wiley
Stantcheva, T., Herbst, E. (2004). A&A, 423, 241
Tafalla, M., Myers, P. C., Caselli, P., Walmsley, C. M. (2004). A&A, 416, 191
Takahashi, J., Williams, D. A. (2000). MNRAS, 314, 273
Teixeira, T. C., Emerson, J. P., Palumbo, M. E. (1998). A&A, 330, 711
Thi, W.-F., van Zadelhoff, G.-J., van Dishoeck, E. F. (2004). A&A, 425, 955
Tielens, A. G. G. M. (2005). *The Physics and Chemistry of the Interstellar Medium*, in press
Tielens, A. G. G. M., Allamandola, L. J. (1987). In: *Interstellar processes*, p. 397 (Dordrecht, D. Reidel Publishing Co.)
Tielens, A. G. G. M., Hagen, W. (1982). A&A, 114, 245
Turner, B. E. (1990). ApJ, 362, L29
van der Tak, F. F. S., Schilke, P., Müller, H. S. P., Lis, D. C., Phillips, T. G., Gerin, M., Roueff, E. (2002). A&A, 388, L53
van der Tak, F. F. S., van Dishoeck, E. F. (2000). A&A, 358, L79
van Dishoeck, E. F. (2004). ARA&A, 42, 119
van Dishoeck, E. F., Blake, G. A. (1998). ARA&A, 36, 317
van Dishoeck, E. F., Blake, G. A., Draine, B. T., Lunine, J. I. (1993). In: *Protostars and planets III*, p. 163
van Dishoeck, E. F., Thi, W.-F., van Zadelhoff, G.-J. (2003). A&A, 400, L1
van Zadelhoff, G.-J., Aikawa, Y., Hogerheijde, M. R., van Dishoeck, E. F. (2003). A&A, 397, 789
Vastel, C., Phillips, T. G., Yoshida, H. (2004). ApJ, 606, L127
Visser, A. E., Richer, J. S., Chandler, C. J. (2002). AJ, 124, 2756
Viti, S., Collings, M. P., Dever, J. W., McCoustra, M. R. S., Williams, D. A. (2004). MNRAS, 354, 1141
Wakelam, V., Caselli, P., Ceccarelli, C., Herbst, E., Castets, A. (2004). A&A, 422, 159
Walmsley, C. M., Flower, D. R., Pineau des Forêts, G. (2004). A&A, 418, 1035

Ward-Thompson, D., Motte, F., André, P. (1999). MNRAS, 305, 143
Watson, W. D. (1974). ApJ, 188, 35
Weingartner, J. C., Draine, B. T. (2001). ApJ, 563, 842
Willacy, K., Langer, W. D. (2000). ApJ, 544, 903
Willacy, K., Millar, T. J. (1998). MNRAS, 298, 562
Williams, D. A., Williams, D. E., Clary, D., et al. (2000). In: *Molecular hydrogen in space*, ed by F. Combes, and G. Pineau des Forêts, p.99 (Cambridge University Press)
Williams, J. P., Bergin, E. A., Caselli, P., Myers, P. C., Plume, R. (1998). ApJ, 503, 689
Wyrowski, F., Hofner, P., Schilke, P., Walmsley, C. M., Wilner, D. J., Wink, J. E. (1997). A&A, 320, L17
Woon, D. E. (2002). ApJ, 569, 541
Wright, M. C. H., Plambeck, R. L., Wilner, D. J. (1996). ApJ, 469, 216

Stan Kurtz, Melvin Hoare and John Richer

The CAUP Gang

# PROTOSTELLAR JETS: A HIGH ANGULAR RESOLUTION PERSPECTIVE

Francesca Bacciotti

*INAF - Osservatorio Astrofisico di Arcetri - Largo E. Fermi, 5, I-50125 Florence, Italy*

fran@arcetri.astro.it

**Abstract** In the last few years new investigation techniques have allowed us to study in deep the spectacular phenomenon of protostellar jets, and to test the validity of the proposed models for their launch and propagation. In this contribution the current knowledge on the subject is reviewed, with a special emphasis on the recent achievements obtained thanks to observations at high angular resolution, like those performed at subarcsecond scales with the Hubble Space Telescope and with Adaptive Optics techniques. These results have made us able to define more clearly the morphology, kinematics, excitation of the flows on small scales, and, in turn, to derive stringent constraints for the physical processes at work. The novel information, collected very recently, puts us in a very good position to plan rich and fruitful observative campaigns with the instruments being developed for astronomy at IR, submm amd mm wavelengths, in interferometric or single-telescope modes.

## Introduction

The evolution of objects in the Universe is often dominated by mass exchange between the object itself and the surrounding environment. One of the best studied examples is the exchange of mass between a star and its parent cloud during the formation phase. In the infancy of a star, in fact, complex and interesting phenomena occur, like the collapse of the parental core, the mass accretion from the circumstellar disk, and the generation of outflows.

Although the physical ingredients that play a role in the star formation game are known (gravity, opacity to radiation, magnetic field, rotation...), there are aspects of the standard formation theory that have posed for a long time problems of difficult solution. Two issues are of particular interest: (i) which is the process that stops the growth of the central star, although there is still plenty of formation material to use around; (ii) given the measured rotation velocities of a molecular cloud, no star could form at the final mass and size if the angu-

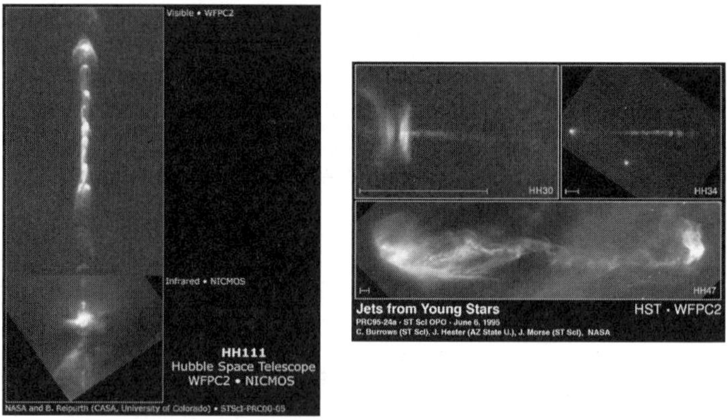


*Figure 1.* Supersonic Jets associated with young stellar objects, imaged in the light of atomic emission lines at high angular resolution (0.″1) with instruments on-board the Hubble Space Telescope.

lar momentum is not extracted someway from the system. How is this excess angular momentum removed ?

A solution to both problems started to emerge when scientists realized that almost all forming stars are associated to impressive mass outflows. These can take the form of large molecular flows, relatively slow and moderately collimated, or extremely energetic, fast and highly collimated atomic jets (Reipurth & Bally 2001, Eislöffel et al. 2000). The latters (see Fig. 1) are called 'Herbig-Haro' (or 'HH') jets, from the name of the two scientists that first identified these peculiar nebulosities. The newly discovered outflows constituted a possible immediate solution to the first question: outflow can clear the environment around a young star, reducing the quantity of material that travels towards and finally lands onto it.

At the beginning, the fact that where one was looking for *infall*, invariably *outflow* signatures were found, generated some surprise in the community. Quite soon, however, several observative evidences and also theoretical modelling indicated that accretion and ejection of matter are phenomena deeply tied together. For example, all jets are associated with stars possessing an accretion disk. In addition, jets are more powerful in the first stages of the formation phase, when accretion also proceeds at a fast pace. Jets then gradually fade out and disappear as accretion slows down and the star approaches the main sequence. Within this accretion/ejection framework, the most widely accepted theoretical scenario for the generation of jets invokes the contemporaneous action of magnetic field and rotation. As it will be described below, the main attractive of this model resides in the fact that the winds can carry away most of the excess angular momentum. Therefore, jets can also be the answer to the second big question posed by star formation, which makes them an essential ingredient of the whole process.

The present chapter includes first a review of the general properties of protostellar jets (Sect. 1). The attention will then be focused on the recent results obtained at high angular resolution, and the implications they have on jet modelling (Sections 2 and 3). Finally, the achievements expected from the new optical, NIR and sub-mm interferometers will be briefly outlined (Sect 4).

# Properties of stellar jets

Collimated outflows are associated to single and multiple Class 0-I-II sources, in a wide range of masses. They appear since the very beginning of the formation process and often have two lobes that propagate in opposite directions perpendicular to the plane of the circumstellar disk. The jet length may vary from a few hundreds AU to several parsecs, while typical widths are of $\sim$ 200 AU, with an opening angle of only a few degrees. Archetypical cases are the HH 34 and HH 111 jets in Orion, and the HH 30 jet in Taurus, see Fig. 1.

**Jet structure.** Jets present a knotty structure along the beam, which is accompayined by several giant 'bow-shocks' that mark the location of the impact onto the surrounding environment, and also can indicate episodes of enhanced mass ejection in the past history of the star. The distinctive feature of these nebulae is their peculiar emission. Stellar jets, in fact, emit strong spectral *lines*, in a large interval of wavelengths, but almost no continuum. The emission is particularly strong in optical and Near Infrared (NIR) forbidden lines of neutral and ionised species, like $O^0$, $S^+$, $N^+$, $Fe^+$, C, $Ca^+$, as well as in permitted lines of Hydrogen, Helium and $Mg^+$. (Reipurth & Bally 2001). Highly collimated jets are also seen in molecular lines of $H_2$, SiO, CO (Eislöffel et al. 2000.) Occasionally the base of the jet is seen in free-free radio continuum, and X-ray emission has been detected at the base of the flow as well as in the big bow shocks.

Both in the beam and in the terminal 'bows' the lines form most likely behind shock fronts that develop during the propagation of the material (see Hartigan 2003). In fact, typical Doppler shifts witness velocities in excess of of 100 - 200 km $s^{-1}$, while the gas temperature corresponding to the observed lines is about $1 - 2\ 10^4$ K. These data show that the jets are highly supersonic (Mach number $v_{\rm jet}/c_s \sim$ 20-30), which implies that every small variation in either the amplitude or the direction of the velocity vectors generate a shock front in the beam that heats and excites the gas.

The many spectral lines emitted by jets are indeed a mine of precious informations for astronomers. Not only the jet velocity can be measured from them, but also a variety of important physical quantities, using specialised diagnostic techniques that compare the measured line ratios with the ratios predicted by radiative models. The list includes the dust extinction, the electron density $n_e$, the electron temperature $T_e$, the hydrogen ionization fraction $x_e$, some elemental abundances, etc. (Bacciotti & Eislöffel 1999; Bacciotti 2002; Nisini et al. 2002; Pesenti et al. 2003; Hartigan et al. 2004). Particularly important are the derivations of the *total density* $n_H$, since all the models for the flow dynamics are highly dependent on this parameter (see the contribution by L. Podio). The realization that stellar jets are only partially ionised (typically $0.02 < x_e < 0.6$) has provided new, more accurate estimates of the density, and jets have been found to be heavier than previously known (on large scales, $n_H$ ranges from $10^3$ to $10^5$ cm$^{-3}$). This, in turn, implies that jets can interact strongly with the environment. For example, the linear momentum transported by the central collimated atomic jet is about $10^{-5}$ M$_\odot$ yr$^{-1}$ km s$^{-1}$. Therefore, jets are in principle capable to drive the coaxial wider but slower molecular ouflow, although the mechanism by which the momentum is transferred is not yet fully understood. Since molecular outflows are thought to be constituted by ambient material set in motion by the passage of the jet giant bow shocks, the central jet 'spine' appears indeed as the major agent for clearing the star's environment, as indicated in the introduction.

**Jet models.** In the following, a few basic physical concepts driving the theoretical research in this field are highlighted.

Regarding the jet propagation, most theoretical studies concentrated on the formation of the knots, either invoking the action of Kelvin-Helmoltz instabilities or the effects of variability in the ejection mechanism (see, e.g. Stone & Hardee 2000). It now appears that the nodular structure in the beam is hardly reproduced by the simulations if some sort of pulsed ejection is not introduced. The majority of the models for the jet launch consider steady magneto-hydrodynamics (see below), and so they cannot account for pulsation. Nevertheless, a viable mechanism producing ejection variability has to be found. In this respect, a number of interesting time-dependent models, that include magnetic instabilities at the star/disk interface have recently been developed (e.g. Matt et al. 2002; Ouyed et al. 2003). Laboratory experiments for the jet propagation have also been set up recently (see e.g., Gardiner 2003), that can give insights in peculiar features of the jet propagation, as jet bending, jet collisions, re-excitation events along the beam.

As for what concerns the jet generation, the most widely accepted class of models is based on the so-called magneto-centrifugal mechanism, that was originally developed for extragalactic jets. According to this theoretical scenario, the hourglass shaped magnetic field attached to the star/disk system rotates with an angular speed related to the angular speed of the 'footpoint' of the magnetic field lines (see fig. 2). The combination of magnetic and centrifugal

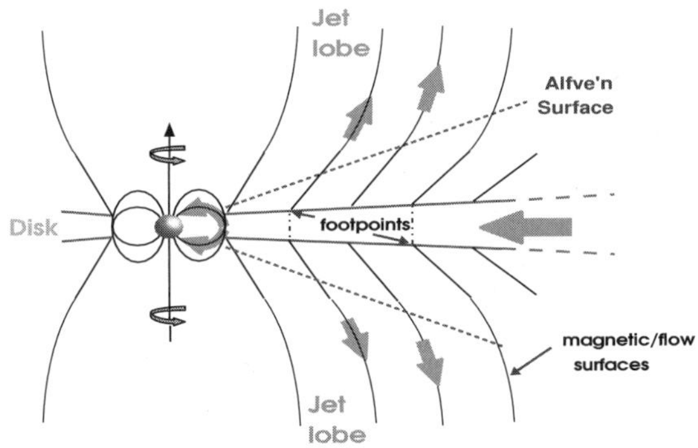


*Figure 2.* Schematic picture of the launching region of a MHD disk wind. Accretion proceeds through the disk toward the star and at the same time outflow lobes are generated by the combined action of centrifugal and magnetic forces.

forces accelerates the partially ionised fluid particles lifted from the star/disk system along the open field lines, in two opposite directions perpendicular to the disk plane. The precise location of the jet origin varies with the model, being either from the star itself (Sauty et al. 2002), from the innermost annulus of the disk, where the star dipolar field is truncated (within 0.1 AU from the star for the *X-wind*, Shu et al. 2000), or from the inner regions of the circumstellar disk (a few AUs from the star for the *disk-wind*, see, e.g. Könıgl & Pudritz 2000; Ferreira 2002). The main attractive of this kind of models is that the magnetic field along which the matter is launched also brakes the rotation of the disk. In this way some angular momentum is extracted from the system, so that the matter in the disk can flow inward and finally accrete onto the central star.

Unfortunately, to test the launching models observationally is not an easy task, due to the small scales scales involved. Almost all models, in fact, predict that the jet properties are defined within the initial 100 AUs from the source, corresponding to less than one arcsecond for the systems closer to Earth. Therefore, to confirm the proposed suggestions we need to observe the 'central engine' with the highest angular resolution available, as described in the next sections.

## High angular resolution observations

Despite many sources of 'classical' jets being heavily embedded, the less powerful flows from more evolved, 'cleared' T Tauri stars provide us with a good window on the central engine. Today this zone begins finally to be resolved in observations from space with the Hubble Space Telescope (HST) and from the ground with the 8m-class telescopes provided with Adaptive Optics

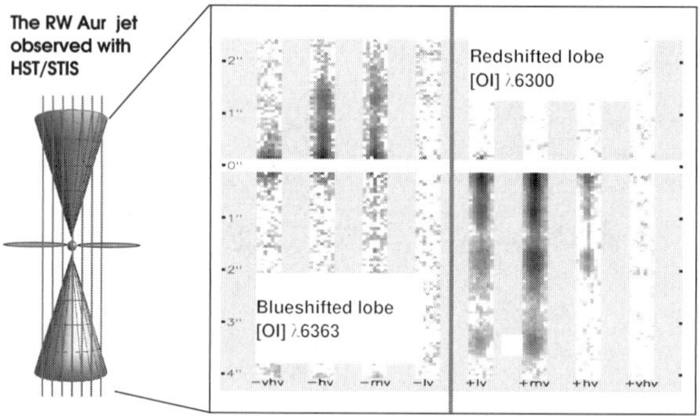


*Figure 3.* 2-D velocity 'channel maps' of the bipolar jet from RW Aur reconstructed from HST/STIS multi-slit optical spectra (adapted from Woitas et al. 2002).

(AO). These systems work at 0."05 - 0."1 angular resolution, which allows astronomers to probe the initial 100-200 AU of the jet.

Observations from the ground with AO has provided $0''.2$ resolution images and spectra of a number of flows, such as the small-scale jets from DG Tau, CW Tau, RW Aur (Dougados et al. 2000, Lavalley et al. 2000). These studies have provided interesting information about the jet diameter and its initial opening angle, the collimation properties, and the values/variation of basic physical parameters along the first portion of the flow (density, temperature, ionization). Generally, the diameter of the flows at the observed height above the disk is about a few tens of AU at the jet base, and collimation is achieved very early, in agreement with the predictions of the magneto-centrifugal models.

This region has been further investigated with the spectrograph STIS on board HST, that worked at $0.''1$ angular resolution in the optical (i.e. 14 AU in Taurus). Although STIS worked with only R $\sim$ 6000, the combination with high angular resolution gave new and important information, as jets could be resolved transversely. For example, we have observed the jets from the TTSs RW Aur and DG Tau with multiple exposures of the $0.''1$ STIS slit, stepping the slit position across the flow every $0.''07$ (see Fig. 3). In this way we have built 3-D cubes of data (2-D spatial, 1-D in velocity) to study in detail the region of interest (Bacciotti et al. 2000; Woitas et al. 2002). One projection of the datacubes gives 2-D images of the jets in different velocity intervals, similarly to the 'channel maps' of radio interferometry. The jets imaged in this way show an onion-like kinematic structure, being more collimated at higher velocities and excitation.

The maps of the line ratios constructed with STIS can then give information about the gas physics. 2-D maps of the various quantities of interest in the different velocity channels have been obtained by Bacciotti 2002. The electron density maps, for example, confirm that $n_e$ is higher closer to the star, the axis,

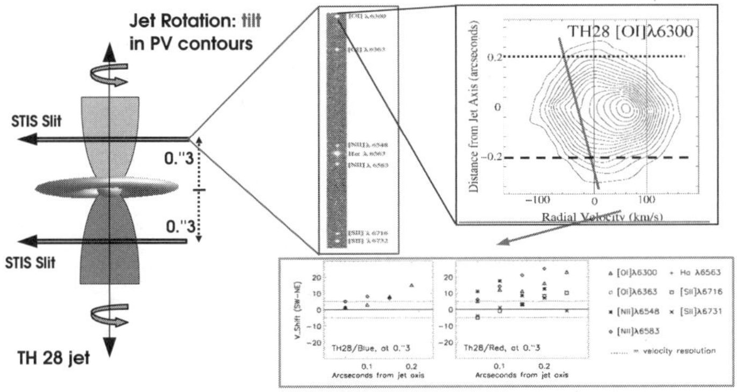


*Figure 4.* HST/STIS observations of the line velocity shifts across the bipolar jet from TH 28 with slits perpendicular to the jet axis (Coffey et al. 2004). The velocity gradient across the jet in the Position-Velocity (PV) diagrams is consistent with rotation around the symmetry axis.

and at higher velocity. At the jet base, one then typically finds 0.01$< x_e <$ 0.4, and total densities up to $10^6$ cm$^{-3}$. In the same region 8 $10^3 < T_e <$ 2 $10^4$ K. These values can be compared with the predictions of magnetic models (see e.g. Garcia et al. 2001; Shang et al. 2002). Finally, from $n_H$, the jet diameter and the de-projected velocity, one can determine the initial mass flux in the jet $\dot{M}_{\rm jet}$, that is found to be of the order of $10^{-7}$ M$_\odot$ yr$^{-1}$, with the colder and slower external layers of the jet contributing the most. According to magneto-centrifugal models, $\dot{M}_{\rm jet}$ should be about 5 - 10% of $\dot{M}_{\rm acc}$, the mass accretion rate through the disk onto the central object. Our observations confirm this prediction in all the cases studied.

**Jet rotation and its impact on the models.** According to the magneto - centrifugal scenario a key element to accelerate the flows is the rotation of the system. Since the mechanism predicts that the jet carries away the excess angular momentum, some trace of rotation should be observable in the outflow immediately above the accelerating region, using high angular resolution instruments that resolve the jet transversely. After the first hints of rotation seen in the HH 212 jet at large distance ($10^4$ AU) from the source by Davis et al. 2000, rotation has been detected in the initial jet channel using STIS/HST. In particular, rotation signatures have been seen in the flows from about ten T Tauri stars, including the bright DG Tau and RW Aur, using different observation techniques and different emission lines (e.g. Bacciotti et al. 2002; Coffey et al. 2004; Woitas et al. 2005). Systematic shifts in the radial velocity of about 6 to 25 $\pm$ 5 km s$^{-1}$ were found in positions of the jet displaced symmetrically with respect to the axis, at 50 - 60 AU from the source and 20 - 30 AU from the axis (see Fig. 4 for the TH 28 jet, observed with slits perpendicular to the flow direction). In this survey, rotation has been detected in all the targets examined, and found to be consistent in the various elements of the system (bipolar lobes, disk (Testi et al. 2002)), and between different datasets.

The above measurements constitute the first global validation of the models proposed for the jet launching. In addition, from the combination of measured toroidal and poloidal velocities one can determine the *"footpoint radius"* of the wind, i.e. the location in the disk from where the observed portion of the rotating wind is launched (cfr. Fig. 2). In all cases examined the observations are consistent with the footpoint being located between 0.5 and 2 AU from the star (Anderson et al. 2003; Coffey et al. 2004; Pesenti et al. 2004; Woitas et al. 2005). This finding appears to support the disk-wind models, that consider an extended region in the disk for the origin of the wind, although X-winds may be present inside, closer to the symmetry axis. Using then mass and angular momentum conservation in the global system, we have verified that a consistent fraction of the excess disk angular momentum (60% to 100% ) can be carried away by the jet. This also indicates that disk viscosity, whose nature is still unidentified, plays anyhow a minor role in the angular momentum balance of the inner regions of the disk.

Finally, we stress that in magneto-centrifugal models, the poloidal and toroidal components of the magnetic field (that cannot be measured directly for stellar jets) are related to the corresponding components of the velocity field. Using this property one finds that at the observed location the toroidal component of the magnetic field is 3 - 4 times larger than the poloidal component (Woitas et al. 2005). A prevalence of the toroidal field component at 50 - 100 AU from the star is indeed predicted by the models, that attribute to a magnetic 'hoop stress', the collimation of the flow (Königl & Pudritz 2000).

## Optical, NIR and Sub-mm Interferometry

Although observing with $0.''1$ resolution has provided precious information about the launch mechanism, the real 'core' of the acceleration engine lies below the so-called 'Alfvén surface' (cfr. Fig. 2), located at a height of a few AUs above the disk. This region will only be approached with Optical/NIR interferometry. Two such big interferometric facilities are being prepared at the time of writing: the VLT Interferometer (VLTI) at ESO-Paranal, and the Large Binocular Telescope (LBT) at Mount Graham in Arizona.

The VLTI is the interferometer obtained by connecting in various combinations the four 8-m telescopes and the smaller auxiliary telescopes at ESO-Paranal. The focal plane instrument, the spectrograph AMBER, combines three beams coming from the telescopes, corrected for atmospheric seeing with AO. AMBER operates at the NIR wavelength range, in three different spectral resolution modes, and will provide an angular resolution of a few milliarcseconds. This will allow us to finally access the innermost part of the flow down to 0.1 AU, where the jet acceleration takes place. The LBT Interferometer is instead formed by the combination of two 8.3m telescopes mounted together in a gigantic binocule, to achieve a total collecting area of a 23m telescope ( Herbst 2003). LBT is the result of a joint collaboration between the Arcetri Observatory (INAF) and several German and US organisations. This instrument will also start the operations working in the the NIR but a future extension to the Optical is contemplated. Although the limited length of the binocule arms provides an angular resolution smaller than VLTI, the instrument has a simpler design for the interferometric mode, which traduces in a mucher higher efficiency. Either way, one should keep in mind that the interferometric observations will not give directly a flow image or spectrum, but instead provide the values of the Fourier transform of the signal (called 'visibility') on a number of locations within the 'conjugate plane' of the spatial frequencies. This plane is sampled during Earth rotation by the 'tracks' drawn by the different baselines, i.e. the lines that join two elements of the interferometer. >From the value observed, with an Inverse Fourier Transform algorithm, the real 'image' of the observed target is then reconstructed. The better the coverage, the higher the quality of the resulting image. Since, however, the coverage will in practice be quite partial, models of the expected emission are needed to interpret the observations (Bacciotti et al. 2003).

The other large interferometric facility being planned for observations at the sub-mm wavelength range is ALMA, an array of 64 12m antennas to be built in the Atacama desert in Chili, not very far from the VLTI site (see L. Testi, this book). With ALMA we will be able to test for the first time the rotation of southern disks and molecular outflows, and also to attack one of the most challenging but urgent scientific issues in the physics of jets, that is the measurement of the magnetic field in the flows.

## Conclusions: universality of the accretion/ejection engine

The results collected at high angular resolution in the last few years all point toward a confirmation of a magneto-centrifugal scenario. The proposed models, developed typically for a 1 $M_\odot$ star, are actually not strongly dependent on the mass of the central object. We may then speculate that the same processes can apply to young star of all masses, and look for observational confirmations. The existence of winds in Brown Dwarfs ($M_\star \leq 0.075\ M_\odot$) is suggested by the detection of strong forbidden lines in the spectra of some of these stars ( Comeron et al. 2003). It is therefore interesting to test if such winds have properties consistent with the extension of the accretion/ejection principle to the lower end of the stellar mass spectrum. On the other hand, we may want to know if the generation of stellar jets follows a similar pattern also in high mass ($M \geq 10\ M_\odot$) stars (see R. Cesaroni, this book), for which the picture is complicated by the huge emission of ionising photons, that alter the properties of the circumstellar medium. Observational campaigns are being planned at these aims, to finally confirm the universality of the jet generation scenario.

## Acknowledgments

I gratefully acknowledge the organisers for their invitation.

## References

Anderson, J. M., Li, Z.-Y., Krasnopolsky, R., Blandford, R.(2003).*ApJ* 590:L107

Bacciotti, F. & Eislöffel, J.(1999).*A&A* 342:717

Bacciotti, F., Mundt, R., T. P. Ray, Eislöffel, J., Solf, J. & Camenzind, M. (2000).*ApJ* 537:L49

Bacciotti, F.(2002).*RMxAC* 13:8

Bacciotti, F., Ray, T. P., Mundt, R., Eislöffel, J. & Solf, J.(2002).*ApJ* 576, 222

Bacciotti, F., et al. (2003).*Ap&SS* 286:157

Coffey, D., Bacciotti, F., Ray, T. P., Woitas, J. & Eislöffel, J. (2004).*ApJ* 604:758

Comeron, F., Fernández, M., Baraffe, I., Neuhäuser, R., & Kaas, A. (2003).*A&A* 406:1001

Davis, C. J., et al.(2000).*MNRAS* 314:241

Dougados, C., Cabrit, S., Lavalley, C. & Ménard, F.(2000).*A&A* 357:61

Eislöffel, J. et al. (2000).In*Protostar & Planets IV (PPIV), eds. V. Mannings, A. P. Boss, S. S. Russell (Tucson)* 815

Ferreira, J.(2002).In *Star Formation and the Physics of Young Stars eds: Bouvier, J. & Zahn, J.-P.(EDP Sciences)* 3:229

Garcia, P., Ferreira, J., Cabrit, S. & Binette, L. (2001).*A&A* 377:589

Gardiner, T., et al. (2003).*Ap&SS* 287:69

Hartigan, P. (2003).*Ap&SS* 287:111

Hartigan, P., Edwards, S. & Pierson, R.(2004).*ApJ* 609:261

Herbst, T. (2003).*Ap&SS* 286:45

Königl, A. & Pudritz, R.(2000).*PPIV* 759

Lavalley-Fouquet, C., Cabrit, S., & Dougados, C.(2000).*A&A* 356:L41.

Matt, S., et al.(2002).*ApJ* 574:232

Nisini, B., Caratti, O., Garatti, A., Giannini, T. & Lorenzetti, D.(2002).*A&A* 393:1035

Ouyed et al.(2003).*ApJ* 583:292

Pesenti, N., Dougados, C., Cabrit, S., O'Brien, D., Garcia, P. & Ferreira, J.(2003).*A&A* 410:155

Pesenti, N., Dougados C., Cabrit S., Ferreira J., O'Brien D. & Garcia P.(2004).*A&A* 416:L9

Reipurth, B. & Bally, J. (2001).*ARA&A* 39:403

Sauty, C., Trussoni, E. & Tsinganos, K.(2002).*ApJ* 389:1068

Shang, H., Glassgold, A. E., Shu, F. H. & Lizano, S.(2002).*ApJ* 564:853

Shu, F. H., Najita, J. R., Shang, H. & Li, Z.-Y.(2000).*PPIV* 789

Stone, J.M. & Hardee, P.E.(2000).*ApJ* 540:192

Testi, L., Bacciotti, F., Sargent, A. I., Ray, T. P. & Eislöffel, J.(2002).*A&A*394:L31

Woitas, J., Ray, T. P., Bacciotti, F., Davis, C. J. & Eislöffel, J. (2002).*ApJ* 580:336

Woitas, J., Bacciotti, F., Ray, T.P., Marconi, A., Coffey, D. & Eislöffel, J. (2005).*A&A* 432:149

# NON-ISOTHERMAL GRAVOTURBULENT FRAGMENTATION

Anne-Katharina Jappsen [1], Ralf S. Klessen [1], Richard B. Larson [2], Yuexing Li [3,4] and Mordecai-Mark Mac Low [3,4]

[1] *Astrophysikalisches Institut Potsdam, An der Sternwarte 16, D-14482 Potsdam, Germany*

[2] *Department of Astronomy, Yale University, New Haven, CT 06520-8101, USA*

[3] *Department of Astronomy, Columbia University, New York, NY 10027, USA*

[4] *Department of Astrophysics, American Museum of Natural History, 79th Street at Central Park West, New York, NY 10024-5192, USA*

akjappsen@aip.de, rklessen@aip.de, larson@astro.yale.edu, yuexing@amnh.org, mordecai@amnh.org

**Abstract** The thermodynamic state of star-forming gas determines its fragmentation behavior and thus plays a crucial role in determining the stellar initial mass function (IMF). We address the issue by studying the effects of a piecewise polytropic equation of state on the formation of stellar clusters in turbulent, self-gravitating molecular clouds using three-dimensional, smoothed particle hydrodynamics simulations. In these simulations stars form via a process we call gravoturbulent fragmentation, i.e., gravitational fragmentation of turbulent gas. To approximate the results of published predictions of the thermal behavior of collapsing clouds, we increase the polytropic exponent $\gamma$ from 0.7 to 1.1 at a critical density $n_{\rm c}$, which we estimated to be $2.5 \times 10^5\ {\rm cm}^{-3}$. The change of thermodynamic state at $n_{\rm c}$ selects a characteristic mass scale for fragmentation $M_{\rm ch}$, which we relate to the peak of the observed IMF. A simple scaling argument based on the Jeans mass $M_{\rm J}$ at the critical density $n_{\rm c}$ leads to $M_{\rm ch} \propto n_{\rm c}^{-0.95}$. We perform simulations with $4.3 \times 10^4\ {\rm cm}^{-3} < n_{\rm c} < 4.3 \times 10^7\ {\rm cm}^{-3}$ to test this scaling argument. Our simulations qualitatively support this hypothesis, but we find a weaker density dependence of $M_{\rm ch} \propto n_{\rm c}^{-0.5\pm0.1}$. We also investigate the influence of additional environmental parameters on the IMF. We consider variations in the turbulent driving scheme, and consistently find $M_{\rm J}$ is decreasing with increasing $n_{\rm c}$. Our investigation generally supports the idea that the distribution of stellar masses depends on the thermodynamic state of the star-forming gas. The thermodynamic state of interstellar gas is a result of the balance between heating and cooling processes, which in turn are determined by fundamental atomic and molecular physics and by chemical abundances.

## Introduction

One of the fundamental unsolved problems in astronomy is the origin of the stellar mass spectrum, the so-called initial mass function (IMF). Observations suggest that there is a characteristic mass $M_{\rm ch}$ for stars in the solar vicinity. The IMF peaks at this characteristic mass which is typically a few tenths of a solar mass. The IMF has a nearly power-law form for larger masses and declines rapidly towards smaller masses (Scale 1998; Kroupa 2002; Chabrier 2003). Scalo, 1998 Kroupa, 2002 Chabrier, 2003

Although the IMF has been derived from vastly different regions, from the solar vicinity to dense clusters of newly formed stars, the basic features seem to be strikingly universal to all determinations (Kroupa 2001). Initial conditions in star forming regions can vary considerably. If the IMF depends on the initial conditions, there would thus be no reason for it to be universal. Therefore a derivation of the characteristic stellar mass that is based on fundamental atomic and molecular physics would be more consistent. Kroupa, 2001 Larson, 2003 Mac Low and Klessen, 2004 Larson, 1981 Fleck, 1982 Padoan, 1995 Padoan et al., 1997 Klessen et al., 1998; Klessen et al., 2000 Klessen, 2001a Padoan and Nordlund, 2002 Schmeja and Klessen, 2004 Jappsen and Klessen, 2004 Bodenheimer, 1995 Papaloizou and Lin, 1995 Lin and Papaloizou, 1996

There are many ways to approach the formation of stars and star clusters from a theoretical point of view. In particular models that connect stellar birth to the turbulent motions ubiquiteously observed in Galactic molecular clouds have become increasingly popular in recent years. See, e.g., the reviews by Larson (2003) and Mac Low & Klessen (2004). The interplay between turbulent motion and self-gravity of the cloud leads to a process we call gravoturbulent fragmentation: Supersonic turbulence generates strong density fluctuations with gravity taking over in the densest and most massive regions (e.g., Larson 1981; Fleck 1982; Padoan 1995; Padoan et al. 1997; Klessen et al. 1998, 2000; Klessen 2001; Padoan & Nordlund 2002). Once gas clumps become gravitationally unstable, collapse sets in. The central density increases and soon individual or whole clusters of protostellar objects form and grow in mass via accretion from their infalling envelopes. The mass accretion rates vary with time and are strongly influenced by the cluster enviroment (Klessen 2001). Schmeja & Klessen (2004) showed that a sharp peak in the mass accretion occurs shortly after the formation of the protostellar core. There is a positive correlation between the value of the peak accretion rate and the final mass of the protostar.

Protostellar collapse is also accompanied by a substantial loss of specific angular momentum, even in the absence of magnetic fields (Jappsen & Klessen 2004). Still, most of the matter that falls in will assemble in a protostellar disk. It is then transported inward by viscous and possibly gravitational torques (e.g., Bodenheimer 1995; Papaloizou & Lin 1995; Lin & Papaloizou 1996). If low angular momentum material is accreted, the disk is stable and most of the material ends up in the central star. In this case, the disk simply acts as a buffer and smooths eventual accretion spikes. It will not delay or prevent the mass growth of the central star by much. However, if material with large specific angular momentum is accreted, then the mass load onto the disk is likely to be faster than inward transport. The disk grows large and may become gravitationally unstable and fragment. This may lead to the formation of a binary or higher-order multiple system (Bodenheimer et al. 2000; Fromang et al. 2002).

However, most current results are based on models that do not treat thermal physics in detail. Typically, a simple isothermal equation of state (EOS) is used. The true nature of the EOS, thus, remains a major theoretical problem in understanding the fragmentation properties of molecular clouds. Li et al., 2003

Recently Li et al. (2003) conducted a systematic study of the effects of a varying polytropic exponent on gravoturbulent fragmentation. Their results showed that $\gamma$ determines how strongly self-gravitating gas fragments. They found that the degree of fragmentation decreases with increasing polytropic exponent in the range $0.2 < \gamma < 1.4$ although the total amount of mass in collapsed cores appears to remain roughly constant through this range. These findings suggest that the IMF might be quite sensitive to the thermal physics. However in their computations, $\gamma$ was left strictly constant in each case. Here we study the effects of using a piecewise polytropic equation of state with a polytropic exponent that changes at a certain critical density $\rho_c$ from $\gamma_1$ to $\gamma_2$:

$$\begin{aligned} P &= K_1\,\rho^{\gamma_1} \qquad && \rho \le \rho_c \\ P &= K_2\,\rho^{\gamma_2} \qquad && \rho > \rho_c \end{aligned} \tag{1}$$

where $K_1$ and $K_2$ are constants, P is the thermal pressure, and $\rho$ is the gas density. We investigate if a change in $\gamma$ determines the characteristic mass of the gas clump spectrum and thus, possibly, the turn-over mass of the IMF. Mac Low and Ossenkopf, 2000; Ossenkopf et al., 2001; Ossenkopf and Mac Low, 2002 Klessen et al., 2004 Vázquez-Semadeni et al., 2005 Benson and Myers, 1989; Myers et al., 1991

## Thermal properties of star-forming clouds

Gravity in galactic molecular clouds is initially expected to be opposed mainly by a combination of supersonic turbulence and magnetic fields (Mac Low & Klessen 2004). The velocity structure in the clouds is always observed to be dominated by large-scale modes (Mac Low & Ossenkopf 2000; Ossenkopf et al. 2001; Ossenkopf & Mac Low 2002). In order to maintain turbulence for some global dynamical timescales and to compensate for gravitational contraction of the cloud as a whole, kinetic energy input from external sources seems to be required. Star formation then takes place in molecular cloud regions which are characterized by local dissipation of turbulence and loss of magnetic flux, eventually leaving thermal pressure as the main force resisting gravity in the small dense prestellar cloud cores that actually build up the stars (Klessen et al. 2004; Vázquez-Semadeni et al. 2005). In agreement with this expectation, observed prestellar cores typically show a rough balance between gravity and thermal pressure (Benson & Myers 1989; Myers et al. 1991). Therefore the thermal properties of the dense star-forming regions of molecular clouds must play an important role in determining how these clouds collapse and fragment into stars. Koyama and Inutsuka, 2000 Larson, 1985; Larson, 2005; Spaans and Silk, 2000

Observational and theoretical studies of the thermal properties of collapsing clouds both indicate that at densities below $10^{-18}\,\mathrm{g\,cm^{-3}}$, roughly corresponding to a number density of $n = 2.5\times10^{5}\,\mathrm{cm^{-3}}$, the temperature decreases with increasing density. This is due to the strong dependence of molecular cooling rates on density (Koyama & Inutsuka 2000). Therefore, the polytropic exponent is below unity in this density regime. At densities above $10^{-18}\,\mathrm{g\,cm^{-3}}$, the gas becomes thermally coupled to the dust grains, which then control the temperature by far-infrared thermal emission. The balance between compressional heating and thermal cooling by dust causes the temperature to increase again slowly with increasing density. Thus the temperature-density relation can be approximated with $\gamma$ above unity (Larson 1985, 2005;Spaans & Silk 2000). Changing from a value below unity to a value above unity results in a minimum temperature at the critical density. As shown by Li et al. (2003), gas fragments efficiently for $\gamma < 1.0$ and less efficiently for higher $\gamma$. Thus, the Jeans mass at the critical density defines a characteristic mass for fragmentation, which may be related to the peak of the IMF.

## Numerical Approach

Springel et al., 2001 Benz, 1990; Monaghan, 1992 Bate and Burkert, 1997 Bate et al., 1995 Jappsen et al., 2005 To gain insight into how molecular cloud fragmentation and the characteristic stellar mass may depend on the critical density, we perform a series of smoothed particle hydrodynamics (SPH) calculations of the gravitational fragmentation of supersonically turbulent molecular clouds using the parallel code GADGET by Springel et al. (2001). SPH is a Lagrangian method, where the fluid is represented by an ensemble of particles, and flow quantities are obtained by averaging over an appropriate subset of SPH particles; see Benz (1990) and Monaghan (1992). The method is able to resolve large density contrasts as particles are free to move, and so naturally

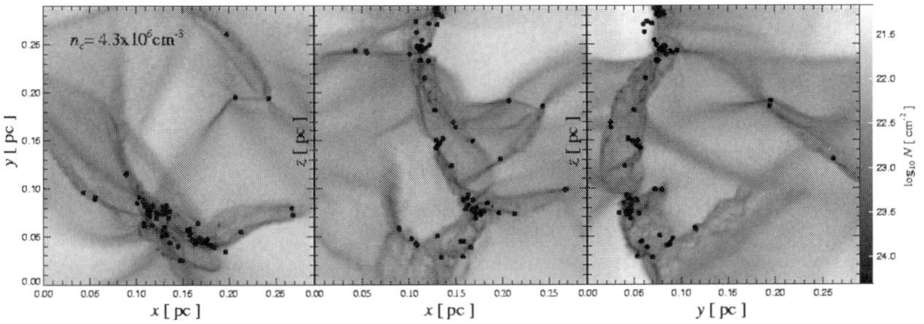


*Figure 1.* Column density distribution of the gas and location of identified protostellar objects (*black circles*) using the model with $n_c = 4.3 \times 10^6 \text{ cm}^{-3}$ where approximately 50% of the gas is accreted. Projections in the xy-, xz-, and yz-plane are shown.

the particle concentration increases in high-density regions. We use the Bate & Burkert (1997) criterion for the resolution limit of our calculations. It is adequate for the problem considered here, where we follow the evolution of highly nonlinear density fluctuations created by supersonic turbulence. We replace the central high-density regions of collapsing gas cores by sink particles (Bate et al. 1995). These particles have the ability to accrete gas from their surroundings while keeping track of the total mass, the linear and angular momentum of the collapsing gas. This enables us to follow the dynamical evolution of the system over many local free-fall timescales (see Fig. 1). Bodenheimer, 1995; Papaloizou and Lin, 1995; Lin and Papaloizou, 1996 Bodenheimer et al., 2000; Fromang et al., 2002

We compute models where the polytropic exponent changes from $\gamma = 0.7$ to $\gamma = 1.1$ for critical densities in the range $4.3 \times 10^4 \text{ cm}^{-3} < n_c < 4.3 \times 10^7 \text{ cm}^{-3}$. Each simulation starts with a uniform density distribution, and turbulence is driven on large scales, with wave numbers $k$ in the range $1 \leq k \leq 2$. We use the same driving field in all four models. The global free-fall timescale is $t_{\text{ff}} \approx 10^5 \text{ yr}$. For further details see Jappsen et al. (2005). Salpeter, 1955 Lada and Lada, 2003 Bonnell et al., 2001 Klessen, 2001b Schmeja and Klessen, 2004 Jappsen and Klessen, 2004

## Dependency of the Characteristic Mass

To illustrate the effects of varying the critical density, we plot in Fig. 2 the resulting mass spectra at different times when the fraction of mass accumulated in protostellar objects has reached approximately stages of 10% and 30%. This range of efficiencies corresponds roughly to the one expected for regions of clustered star formation (Lada & Lada 2003). In the model with a critical density $n_c$ of $4.3 \times 10^4 \text{ cm}^{-3}$ the change in $\gamma$ occurs below the initial mean

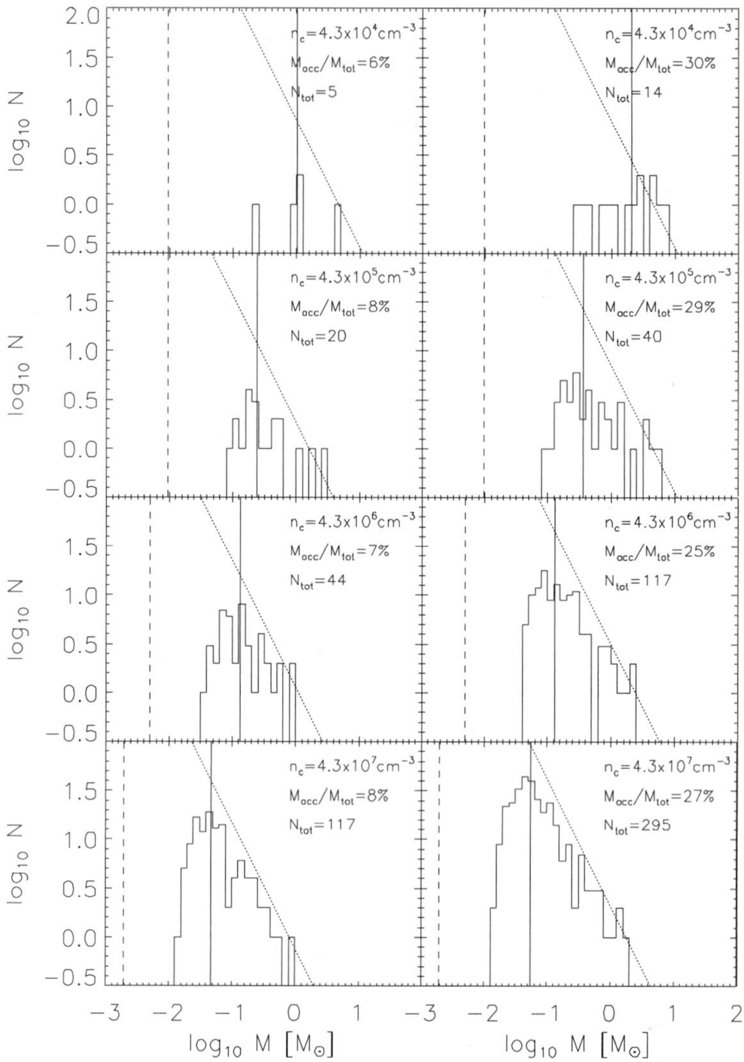


*Figure 2.* Mass spectra of protostellar cores for four models with critical densities in the range $4.3 \times 10^4\ \mathrm{cm}^{-3} < n_c < 4.3 \times 10^7\ \mathrm{cm}^{-3}$. We show two phases of evolution, when about $10\%$ and $30\%$ of the mass has been accreted onto protostars. The *vertical solid line* shows the median mass of the distribution. The *dotted line* serves as a reference to the Salpeter (1955) slope of the observed IMF at high masses. The *dashed line* indicates our mass resolution limit.

density. The mass spectrum shows a flat distribution with only few, but massive protostellar objects. These reach masses up to $10\ M_\odot$ and the minimum mass is about $0.3\ M_\odot$. All other models build up a power-law tail towards high masses. This is due to protostellar accretion processes, as more and more gas gets turned into stars (see also, Bonell et al. 2001; Klessen 2001; Schmeja &

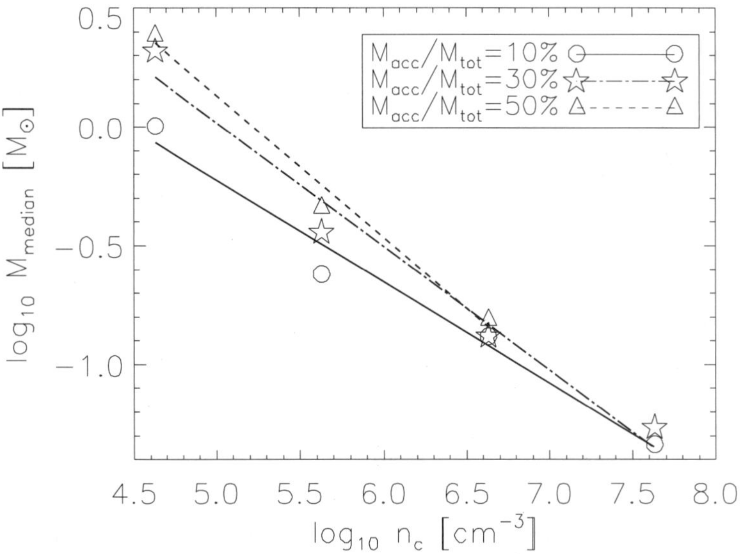


*Figure 3.* Plot of the median mass of the protostellar cores $M_{\rm median}$ versus critical density $n_{\rm c}$. We display results for different ratios of accreted gas mass to total gas mass, $M_{\rm acc}/M_{\rm tot}$ , and fit the data with straight lines. Their slopes take the values $-0.43 \pm 0.05$ (*solid line*), $-0.52 \pm 0.06$ (*dashed-dotted line*), and $-0.60 \pm 0.07$ (*dashed line*), respectively.

Klessen 2004). The mass spectrum becomes more peaked for higher $n_{\rm c}$ and shifts to lower masses.

We find closest correspondence with the observed IMF (Scalo 1998; Kroupa 2002; Chabrier 2003) for a critical density $n_{\rm c}$ of $4.3 \times 10^6$ cm$^{-3}$ and for stages of accretion around 30%. For high masses, the distribution exhibits a (Salpeter 1955) power-law behavior. For masses about the median mass the distribution has a small plateau and then falls off towards smaller masses.

The change of median mass $M_{\rm median}$ with critical density $n_{\rm c}$ is quantified in Fig. 3. As $n_{\rm c}$ increases $M_{\rm median}$ decreases. We fit our data with straight lines. The slopes take values between $-0.4$ and $-0.6$.

## Summary

Using SPH simulations we investigate the influence of a piecewise polytropic EOS on fragmentation of molecular clouds. We study the case where the polytropic index $\gamma$ changes from 0.7 to 1.1 at a critical density $n_{\rm c}$, and consider the range $4.3 \times 10^4\ {\rm cm}^{-3} < n_{\rm c} < 4.3 \times 10^7\ {\rm cm}^{-3}$ around the realistic value to determine the dependence of the mass spectrum on $n_{\rm c}$. Koyama and Inutsuka, 2000

Our investigation supports the idea that the distribution of stellar masses depends, at least in part, on the thermodynamic state of the star-forming gas. If there is a low-density regime in molecular clouds where temperature $T$ sinks with increasing density $\rho$, followed by a higher-density phase where $T$ increases with $\rho$, fragmentation seems likely to be favored at the transition density where the temperature reaches a minimum. This defines a characteristic mass scale. The thermodynamic state of interstellar gas is a result of the balance between heating and cooling processes, which in turn are determined by fundamental atomic and molecular physics and by chemical abundances. The derivation of a characteristic stellar mass can thus be based on quantities and constants that depend solely on the chemical abundances in a molecular cloud.

A simple scaling argument based on the Jeans mass $M_{\rm J}$ at the critical density $n_{\rm c}$ leads to $M_{\rm J} \propto n_{\rm c}^{-0.95}$ (see Jappsen et al. 2004). If there is a close relation between the average Jeans mass and the gravoturbulent fragmentation spectrum, a similar relation should hold for the characteristic mass $M_{\rm ch}$ of protostellar cores. Our simulations qualitatively support this hypothesis, however, with the weaker density dependency $M_{\rm ch} \propto n_{\rm c}^{-0.5\pm0.1}$.

The density at which $\gamma$ changes from below unity to above unity selects a characteristic mass scale. Consequently, the peak of the resulting mass spectrum decreases with increasing critical density. This spectrum not only shows a pronounced peak but also a powerlaw tail towards higher masses. Its behavior is thus similar to the observed IMF.

Altogether, supersonic turbulence in self-gravitating molecular gas generates a complex network of interacting filaments. The overall density distribution is highly inhomogeneous. Turbulent compression sweeps up gas in some parts of the cloud, but other regions become rarefied. The fragmentation behavior of the cloud and its ability to form stars depend strongly on the EOS. If collapse sets in, the final mass of a fragment depends not only on the local Jeans criterion, but also on additional processes. For example, protostars grow in mass by accretion from their surrounding material. In turbulent clouds the properties of the gas reservoir are continuously changing. In a dense cluster environment, furthermore, protostars may interact with each other, leading to ejection or mass exchange. These dynamical factors modify the resulting mass spectrum, and may explain why the characteristic stellar mass depends on the EOS more weakly than expected.

We also studied the effects of different turbulent driving fields and of a smaller driving scale. For different realizations of statistically identical large-scale turbulent velocity fields, we consistently find that the characteristic mass decreases with increasing critical mass. However, there are considerable variations. The influence of the natural stochastic fluctuations in the turbulent flow on the resulting median mass is almost as pronounced as the changes of the thermal properties of the gas. Also when inserting turbulent energy at small wavelengths, we see the peak of the mass spectrum decrease with increasing critical density.

The current study using a piecewise polytropic EOS can only serve as a first step. Future work will need to consider a realistic chemical network and radiation transfer processes in gas of varying abundances.

## Acknowledgments

We acknowledge interesting and stimulating discussions with S. Glover, M. Spaans and S. Kitsionas. A. K. J. thanks the AMNH for its warm hospitality and the Kade Foundation for support of her visits there. A. K. J. and R. S. K. acknowledge support from the Emmy Noether Program of the Deutsche Forschungsgemeinschaft (grant no. KL1358/1). Y. L. and M.-M. M. L. acknowledge partial support by NASA grants NAG5-13028 and NAG5-10103, and by NSF grant AST03-07793.

## References

Bate, M. R., Bonnell, I. A., and Price, N. M. (1995). *MNRAS*, 277:362–376.

Bate, M. R. and Burkert, A. (1997). *MNRAS*, 288:1060–1072.

Benson, P. J. and Myers, P. C. (1989). *ApJS*, 71:89–108.

Benz, W. (1990). In *Numerical Modelling of Nonlinear Stellar Pulsations Problems and Prospects, ed. J. R. Buchler (Dordrecht: Kluwer)*, page 269.

Bodenheimer, P. (1995). *ARA&A*, 33:199–238.

Bodenheimer, P., Burkert, A., Klein, R. I., and Boss, A. P. (2000). *Protostars and Planets IV, ed. V. Mannings, A. P. Boss & S. S. Russell (Tucson: Univ. of Arizona Press)*, page 675.

Bonnell, I. A., Clarke, C. J., Bate, M. R., and Pringle, J. E. (2001). *MNRAS*, 324:573–579.

Chabrier, G. (2003). *PASP*, 115:763–795.

Fleck, R. C. (1982). *MNRAS*, 201:551–559.

Fromang, S., Terquem, C., and Balbus, S. A. (2002). *MNRAS*, 329:18–28.

Jappsen, A.-K. and Klessen, R. S. (2004). *å*, 423:1–12.

Jappsen, A.-K., Klessen, R. S., Larson, R. B., Li, Y., Klessen, R. S., and Mac Low, M.-M. (2005). *å, in press, astro-ph/0410351*.

Klessen, R. S. (2001a). *ApJ*, 556:837–846.

Klessen, R. S. (2001b). *ApJ*, 550:L77–L80.

Klessen, R. S., Ballesteros-Paredes, J., Vázquez-Semadeni, E., and Duran-Rojas, C. (2004). *ApJ*, 620:786–794.

Klessen, R. S., Burkert, A., and Bate, M. R. (1998). *ApJ*, 501:L205.

Klessen, R. S., Heitsch, F., and Mac Low, M.-M. (2000). *ApJ*, 535:887–906.

Koyama, H. and Inutsuka, S. (2000). *ApJ*, 532:980–993.

Kroupa, P. (2001). *MNRAS*, 322:231–246.

Kroupa, P. (2002). *Science*, 295:82–91.

Lada, C. J. and Lada, E. A. (2003). *ARA&A*, 41:57–115.

Larson, R. B. (1981). *MNRAS*, 194:809–826.

Larson, R. B. (1985). *MNRAS*, 214:379–398.

Larson, R. B. (2003). *Rep. Prog. Phys.*, 66:1651–1697.

Larson, R. B. (2005). *MNRAS, submitted, astro-ph/0412357*.

Li, Y., Klessen, R. S., and Mac Low, M.-M. (2003). *ApJ*, 592:975–985.

Lin, D. N. C. and Papaloizou, J. C. B. (1996). *ARA&A*, 34:703–748.

Mac Low, M.-M. and Klessen, R. S. (2004). *Rev. Mod. Phys.*, 76:125–194.

Mac Low, M.-M. and Ossenkopf, V. (2000). *å*, 353:339–348.

Monaghan, J. J. (1992). *ARA&A*, 30:543–574.

Myers, P. C., Fuller, G. A., Goodman, A. A., and Benson, P. J. (1991). *ApJ*, 376:561–572.

Ossenkopf, V., Klessen, R. S., and Heitsch, F. (2001). *å*, 379:1005–1016.

Ossenkopf, V. and Mac Low, M.-M. (2002). *å*, 390:307–326.

Padoan, P. (1995). *MNRAS*, 277:377–388.

Padoan, P. and Nordlund, Å. (2002). *ApJ*, 576:870–879.

Padoan, P., Nordlund, A., and Jones, B. J. T. (1997). *MNRAS*, 288:145–152.

Papaloizou, J. C. B. and Lin, D. N. C. (1995). *ARA&A*, 33:505–540.

Salpeter, E. E. (1955). *ApJ*, 121:161–+.

Scalo, J. (1998). In *ASP Conf. Ser. 142: The Stellar Initial Mass Function (38th Herstmonceux Conference), ed. G. Gilmore & D. Howell (San Francisco: Astron. Soc. Pac.)*, page 201.

Schmeja, S. and Klessen, R. S. (2004). *å*, 419:405–417.

Spaans, M. and Silk, J. (2000). *ApJ*, 538:115–120.

Springel, V., Yoshida, N., and White, S. D. M. (2001). *New Astronomy*, 6:79–117.

Vázquez-Semadeni, E., Kim, J., Shadmehri, M., and Ballesteros-Paredes, J. (2005). *ApJ*, 618:344–359.

Jonathan Tan, Phil Myers, Malcolm Walmsley and Mario Tafalla

# THE BIRTH OF MASSIVE STARS AND STAR CLUSTERS

Jonathan C. Tan
*Institute of Astronomy, Dept. of Physics, ETH Zürich, 8093 Zürich, Switzerland*
jt@phys.ethz.ch

**Abstract** In the present-day universe, it appears that most, and perhaps all, massive stars are born in star clusters. It also appears that all star clusters contain stars drawn from an approximately universal initial mass function, so that almost all rich young star clusters contain massive stars. In this review I discuss the physical processes associated with both massive star formation and with star cluster formation. First I summarize the observed properties of star-forming gas clumps, then address the following questions. How do these clumps emerge from giant molecular clouds? In these clustered environments, how do individual stars form and gain mass? Can a forming star cluster be treated as an equilibrium system or is this process too rapid for equilibrium to be established? How does feedback affect the formation process?

## Introduction

Star clusters[1] are the fundamental units of star formation in galaxies. Most Galactic stars are born in clusters (Lada & Lada 2003): their figure 2 implies that equal numbers of stars are forming in each logarithmic interval of cluster mass, for cluster masses from $\sim 50 - 1000\ M_{\odot}$. There is a dearth of star formation in clusters below $50\ M_{\odot}$. The sample of Lada & Lada (2003) is too small to constrain the initial cluster mass function beyond $\sim 1000\ M_{\odot}$. Hubble Space Telescope observations have probed this range in external galaxies, finding a continuation of the mass function slope (Larsen 2002). For the dwarf starburst galaxy NGC 5253, Tremonti et al. (2001) have proposed a model in which all star formation occurs in clusters, which then dissolve on timescales of $\sim 10$ Myr to create the sources of the observed diffuse UV light.

The initial mass function of stars in clusters appears largely invariant (Kroupa 2002) so that almost all relatively massive clusters will contain at least a few

[1] I define a star cluster as a group of stars that forms together from a gravitationally bound gas *clump*.

high-mass stars. Thus a significant fraction of all stars form in proximity to massive stars, and may be affected by their strong feedback.

Locally, essentially all massive stars form in clusters (de Wit et al. 2005), so high-mass star formation seems to require an environment that will also produce a large number and mass of low-mass stars. In the present-day universe, massive star formation and star cluster formation are one and the same process.

It is clear that an understanding of massive star and star cluster formation is important to many areas of astrophysics, from galaxy evolution to planet formation.

## Overview of physical properties

Figure 1 shows the masses, $M$, and mean surface densities, $\Sigma = M/(\pi R^2)$, of star clusters and interstellar gas clouds. For convenience $\Sigma = 1$ g cm$^{-2}$ corresponds to $4800\ M_\odot$ pc$^{-2}$, $N_{\rm H} = 4.3 \times 10^{23}$ cm$^{-2}$ and $A_V = 200$ mag, for the local gas to dust ratio. Contours of constant radial size, $R$, and hydrogen number density, $n_{\rm H} = \rho/\mu = 3M/(4\pi R^3 \mu)$, where $\mu = 2.34 \times 10^{-24}$ g is the mean mass per H, are indicated. The density contours also correspond to free-fall timescales, $t_{\rm ff} = \sqrt{3\pi/(32G\rho)} = 1.38 \times 10^6 (n_{\rm H}/10^3\ {\rm cm}^{-3})^{-1/2}$ yr. For a virialized cloud with virial parameter $\alpha_{\rm vir} \simeq 1$ (Bertoldi & McKee 1992) the signal crossing or dynamical timescale is $t_{\rm dyn} = 2.0 t_{\rm ff}$.

The presence of molecules allows interstellar gas to cool to low temperatures, $\sim 10 - 20$ K, effectively removing thermal pressure support. To survive the destructive local interstellar FUV radiation field requires a total column of $N_{\rm H} = (0.4, 2.8) \times 10^{21}$ cm$^{-2}$ for $H_2$ and CO, respectively (McKee 1999). Giant molecular clouds (GMCs) have an approximately constant column of $N_{\rm H} = (1.5 \pm 0.3) \times 10^{22}$ cm$^{-2}$ and typical masses $\sim 10^5 - 10^6\ M_\odot$ (Solomon et al. 1987). A sample of local ($d \lesssim 3$ kpc) infrared-dark clouds (IRDCs), discussed in §3, have masses ranging from several hundred to $\sim 10^4\ M_\odot$ and $\Sigma \sim 0.1$ g cm$^{-2}$ (Kirkland & Tan, in prep.), about a factor of 3 greater than the mean value of GMCs. The massive star forming clumps observed in the sub-mm by Mueller et al. (2002) have similar masses, but surface densities typically a factor of five greater still. More revealed star clusters, such as the Orion Nebula Cluster, have similar properties. More massive and higher surface density clusters are rare, but can be found in the Galactic center, e.g. the Arches and Quintuplet clusters (e.g. Kim et al. 2000). The most massive young clusters, so-called super star clusters, are often found in starburst environments, such as the Antennae galaxies, and in some dwarf galaxies, e.g. NGC 5253 (Turner et al. 2000) and NGC 1569 (Gilbert & Graham 2003).

All high-mass star-forming systems appear to be at about a constant density of $n_{\rm H} \sim 2 \times 10^5$ cm$^{-3}$, corresponding to $t_{\rm ff} \sim 1 \times 10^5$ yr. This is about the same as the density at which the cooling rate is a maximum (Larson 2005), and thus gravitational collapse is easiest. A spherical self-gravitating cloud in hydrostatic equilibrium with mean surface density $\Sigma$ and density profile $\rho \propto r^{-k_\rho}$ with $k_\rho = 1.5$, similar to observed clumps, has a mean pressure of $4.3 \times 10^8 \Sigma^2$ K cm$^{-3}$ (McKee & Tan 2003). Massive stars and star clusters appear to form under pressures $\gtrsim 3 \times 10^7$ K cm$^{-3}$, much higher than that of the local diffuse ISM, i.e. $2.8 \times 10^4$ K cm$^{-3}$ (Boulares & Cox 1990).

## Setting up initial conditions for star cluster formation

What causes a particular region of a GMC to form a star cluster? From Figure 1 we see that the surface density, pressures, and volume densities must increase by at least factors of 10, 100, and 1000, respectively. This occurs in only a small part of the GMC: typically only $\sim$1% of the mass is involved.

Models for the cause of star formation can be divided into two groups: quiescent and triggered. In the former, star formation occurs in the densest, most unstable clumps of the GMC, and these form out of the general gravitational contraction of the entire cloud. This process may be regulated by the decay of turbulence, ambipolar (Mouschovias 1996) or turbulent diffusion of magnetic flux, or heating and ionization (McKee 1989) from newly-formed star clusters. In models of triggered star formation, the star-forming clumps are created by compression of parts of the GMC by external causes, such as: cloud collisions (Scoville et al. 1986; Tan 2000); convergent turbulent flows (e.g. Mac Low & Klessen 2004); or feedback from young stars with ionization (Elmegreen & Lada 1977; Thompson et al. 2004; Deharveng et al. 2005), stellar winds (e.g. Whitworth & Francis 2002), protostellar outflows, radiation pressure, and supernovae (e.g. Palous et al. 1994).

Elmegreen (2004) has noted that the compressions that result from most forms of stellar feedback are probably only efficient within particular GMCs or GMC complexes, i.e. young stars forming in one GMC are unlikely to trigger star formation in another. Oey et al. (2005) claim the age sequence of 3 regions of the W3/W4 complex is evidence for triggered star formation over an approximately 100 pc scale region.

A promising method for addressing the cause of star cluster formation is the study of infrared-dark clouds (IRDCs). These are regions that have surface densities high enough to obscure the Galactic infrared background. Large numbers of IRDCs have been found towards the inner Galaxy with the Midcourse Space Experiment (MSX) (Egan et al. 1998). These are associated with dense molecular gas (Carey et al. 1998; Teyssier, Hennebelle & Pérault 2002). Carey et al. (1998) measured the following physical properties: radii $r \sim 0.2 - 8$ pc, column densities $\Sigma \sim 0.5 - 50\,\mathrm{g\,cm^{-2}}$, densities $n_{\rm H} \gtrsim 2 \times 10^5\,\mathrm{cm^{-3}}$, masses $M \sim 10^3 - 10^5\,M_\odot$ and temperatures $T \lesssim 20$ K. Kirkland & Tan (in prep.) identified a sample of relatively nearby IRDCs from the MSX infrared survey

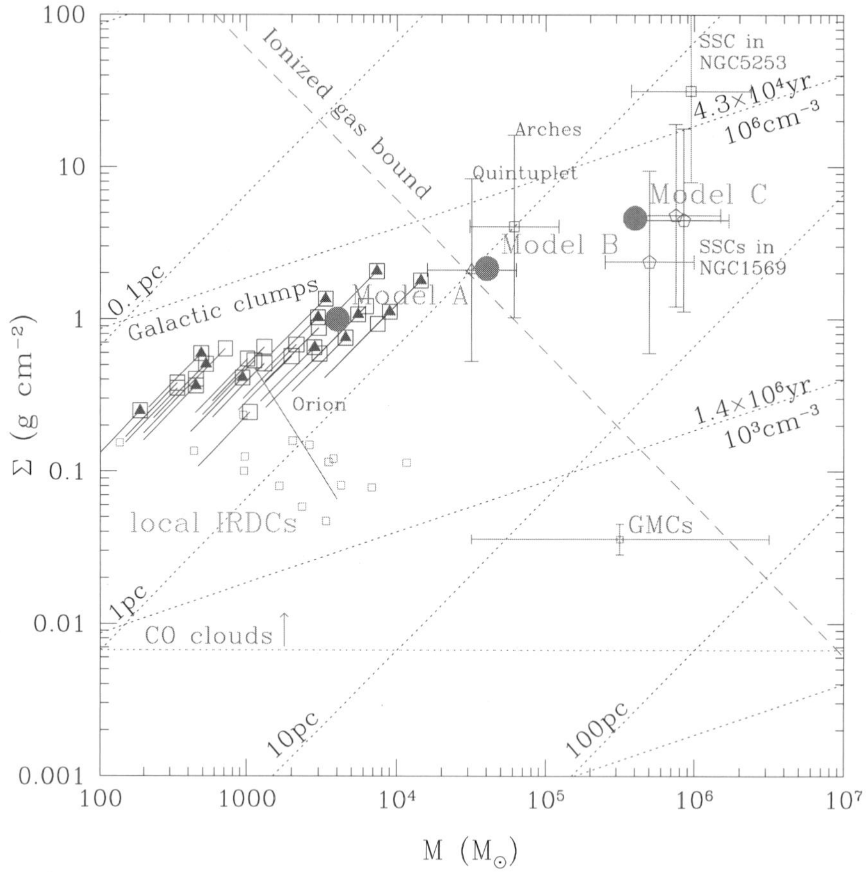


*Figure 1.* Surface density, $\Sigma$, versus mass, $M$, for star clusters and interstellar clouds. Contours of constant radius , $R$, and hydrogen number density, $n_{\rm H}$, or free-fall timescale, $t_{\rm ff}$, are shown with dotted lines. The minimum surface density for CO clouds in the local Galactic UV radiation field is shown, as are typical parameters of GMCs. Dense, cold clumps of GMCs known as infrared dark clouds are shown by small open squares (see text in §3). Large open squares are the star-forming clumps of Mueller et al. (2002): a triangle indicates the clump contains an HII region while the diagonal line from each point shows the effect of uncertain dust opacities on the mass estimate. The Orion Nebula Cluster, allowing for a contribution from gas of 50%, is shown by the solid diagonal line, which traces conditions from the inner to the outer parts of the cluster. Several more massive clusters are also indicated (see §2 for references). The condition for ionized gas to remain bound is indicated by the dashed line. The three solid circles are the conditions of feedback models discussed in §6.

and the Galactic Ring Survey of $^{13}$CO (Simon et al. 2001). Surface densities and masses were estimated with three independent methods: infrared extinction, line strengths of $^{13}$CO, and virial arguments. The dispersion between these methods is a factor of a few due to systematic errors. The average properties of each cloud are shown in Figure 1. They have similar masses to the star clusters and surface densities that are somewhat smaller. Thus they are likely to be representative of the earliest stage of star cluster formation.

From a visual inspection of the sample, Kirkland & Tan (in prep.) find that the morphologies of the IRDCs are more varied, and in particular filamentary, than the star-forming clumps, that are often approximately spherical (Shirley et al. 2003). The line widths are several km/s, which is much greater than the sound speed of gas at $\sim$ 20 K. Thus the initial conditions for star cluster formation probably involve supersonic turbulence, although not necessarily super-Alfvénic turbulence. There are often multiple velocity components. Some IRDCs are relatively isolated from other star-forming regions suggesting that their formation does not require triggering by feedback from young stars.

## How do stars form within clusters?

We have seen that star clusters are born from turbulent gas, i.e. having velocity dispersions much greater than thermal. A basic question is how individual stars form in this environment. In particular do they grow inside quasi-equilibrium gas *cores* that collapse via accretion disks with relatively stable orientations? In this scenario (e.g. Shu, Adams, & Lizano 1987, McKee & Tan 2003) the initial mass of the core helps to determine the final mass of the star, modulo the effects of protostellar feedback. Alternative models involving competitive Bondi-Hoyle accretion (e.g. Bonnell et al. 2001) and direct stellar collisions (Bonnell, Bate, & Zinnecker 1998) have been proposed. These alternative models have been particularly motivated for the case of massive star formation since this occurs in the most crowded regions, radiation pressure feedback from massive stars on dust grains can cause problems for standard accretion scenarios, and the Jeans mass in these high pressure, high density regions is only a fraction of a solar mass.

## Formation of Cores

First consider the formation of cores from a turbulent medium. Ballesteros-Paredes, Klessen, & Vázquez-Semadeni (2003) and Klessen et al. (2005) find that a substantial fraction of "cores" identified in their nonmagnetic SPH simulations of supersonic turbulence appear to be quiescent (i.e. line widths $\leq$ than thermal) and coherent (i.e. line widths are roughly independent of positional offset from the core center), but are in fact dynamic, transient entities. They argue that the inference of hydrostatic equilibrium, e.g. from radial profiles that appear similar to Bonnor-Ebert profiles (e.g. Alves, Lada, & Lada 2001), is not necessarily valid, since such profiles are also possible for dynamically evolving cores. However, it is not clear if these artificial dynamic cores are consistent with the observations of Walsh et al. (2004), which find very small ($\lesssim$ 0.1 km s$^{-1}$) velocity differences between the line centers of high ($n_{\rm H} \sim 4 \times 10^5$ cm$^{-3}$) and low ($n_{\rm H} \sim 2 \times 10^3$ cm$^{-3}$) density traces of starless cores: real cores do not appear to be moving with respect to their envelopes. Estimates of the ages of starless cores (e.g. Crapsi et al. 2005) are uncertain, but have the potential to constrain models of core formation.

The numerical simulations described above are nonmagnetic. Li et al. (2004) and Vázquez-Semadeni et al. (2005) have studied the properties of cores forming from turbulent, magnetized gas. The latter authors find in their periodic, fixed grid, isothermal, ideal MHD, driven turbulence simulations, that: magnetic fields reduce the probability of core formation; in the magnetically subcritical run, a bound core forms that lasts $\sim 5t_{\rm ff}$ (defined at densities $\sim 50$ times the mean), which would be enough for ambipolar diffusion to affect the dynamics; in the moderately supercritical case, where magnetic fields are relatively weaker, bound cores form and then are able to undergo runaway collapse over about $2t_{\rm ff}$, defined at the core's mean density. These results suggest that the initial conditions for star formation are bound cores, and that the stronger the magnetic field, the more chance the cores have to attain a quasi-equilibrium structure. The marginally critical case is probably most relevant if star-forming clumps evolve from regions of GMCs that gradually lose magnetic support. The observations by Crutcher (1999) of the magnetic field strength in dense regions of GMCs imply that these regions are only marginally supercritical and that magnetic fields are important for the dynamics.

Magnetic fields are likely to affect the masses of cores that are present in a given environment. One argument against massive star formation from cores has been that the thermal Jeans mass in the high pressure, high density regions associated with massive star formation is very small. However this argument is irrelevant if massive cores derive most of their pressure support from either magnetic fields or turbulent motions. Observations suggest that the mass function of cores is fairly similar, within large uncertainties, to that of stars and that there are some massive pre-stellar cores (Testi & Sargent 1998; Motte et al. 2001; Li, Goldsmith, & Menten 2003; Beuther & Schilke 2004).

## Accretion to Stars

It is computationally expensive to follow gravitational collapse to the high densities and short timescales associated with protostars and their accretion disks. A common numerical technique is to introduce sink particles in bound regions of high density, which can then accrete gas from their surroundings (Bate, Bonnell, & Price 1995). Bonnell & Bate (2002) modeled star cluster formation with SPH, isothermal, non-periodic, no feedback, nonmagnetic simulations, with initial setup of static gas and sink particles about to undergo global collapse. Stars gained mass via competitive accretion and stellar collisions and the final mass spectrum was similar to the Salpeter mass function. Using similar simulations, except now with an initially turbulent velocity field with no later driving, Bonnell, Vine, & Bate (2004) showed that the most massive star at the end of their calculation had gained mass that was initially very widely distributed. Dobbs, Bonnell, & Clark (2005) found that a massive turbulent core, such as envisaged by McKee & Tan (2003), can fragment into many smaller cores and protostars if the equation of state is isothermal. However, their non-isothermal model suffered much less fragmentation, while the results of Vázquez-Semadeni et al. (2005) suggest that less fragmentation would also occur if magnetic fields are allowed to affect the dynamics. Schmeja & Klessen (2004) simulated star cluster formation with SPH simulations with periodic boundaries, driven turbulence, no magnetic fields, no feedback, sink particle diameters of 560 AU, and an isothermal equation of state, finding highly variable accretion rates for their protostars.

We have seen that SPH simulations, by lacking magnetic fields, probably do not accurately model the fragmentation process of real star-forming regions, particularly with regard to core formation. Another difficulty is that in SPH simulations with sink particles, "stars" acquire most of their mass by competitive Bondi-Hoyle accretion and this process is not adequately resolved. In theory, gas is gravitationally focused by a passing star so that streamlines collide, shock and dissipate their energy. Eulerian grid simulations, including sink particles (Krumholz, McKee, & Klein 2004) and adaptive mesh refinement of small scale structures, have been used to simulate the interaction of sink particles with surrounding turbulent gas: the accretion rate is much smaller than the classical analytic estimate of accretion from a uniform medium (Krumholz, McKee, & Klein 2005a). Stellar feedback should also reduce this accretion rate, particularly to massive stars (e.g. Edgar & Clarke 2004). Thus the importance of Bondi-Hoyle accretion may be grossly over-estimated in SPH simulations.

## Assumptions and Predictions of the McKee-Tan Model

McKee & Tan (2002; 2003) modeled massive star formation by assuming an initial condition that is a massive core in approximate pressure equilibrium with the surrounding protocluster medium, i.e. the star-forming clump. Tan & McKee (2002) modeled star cluster formation by extending this assumption to every star. To derive the pressure in the clump, the system was assumed to be in approximate hydrostatic equilibrium so that the mean pressure is related to the surface density, i.e. $P \sim G\Sigma^2$. How valid are these assumptions?

The mean pressure in the clump sets the overall density normalization of each core and thus its collapse time and accretion rate. The McKee-Tan model allows for deviations from exact pressure equilibrium with the parameter $\phi_P$, although the expectation is that these deviations will be factors of order unity. Although the core is treated as collapsing in isolation, this is also an approximation: McKee & Tan (2003) estimate that during the collapse time the core will interact with an amount of mass similar to the initial core mass, although not all of this will become bound to the core. Thus in reality one would expect for a given initial core mass somewhat greater final stellar masses than under the assumption of isolated cores. The particular density structure of cores assumed by McKee & Tan (2003) is $\rho \propto r^{-k_\rho}$ with $k_\rho = 1.5$ set from observed cores. This choice affects the evolution of the accretion rate during the collapse: $k_\rho < 2$ implies accretion rates accelerate. However, this is a secondary effect compared to the overall normalization of the accretion rate that is set by the external pressure. In any case since the pressure support is nonthermal with significant contributions from turbulent motions, one does not expect a smooth density distribution in the collapsing core, and the accretion rate will show large variations about the mean.

The assumption of approximate pressure equilibrium in the protocluster requires star formation to occur over at least several dynamical timescales, and this is examined in the next section. The basic picture of star cluster formation then involves: a turbulent, self-gravitating gas clump in which bound cores occasionally form (most gas at any given time is not in bound, unstable cores); a core mass function fairly similar to stars, i.e. massive cores form but are rare; an approximate equilibrium of cores with their surroundings; the collapse of cores quite rapidly in one or two free-fall timescales to form stars or binaries; the orbiting of newly-formed stars in the still star-forming clump, but negligible growth via competitive accretion.

Some of the key predictions of the McKee-Tan model are the properties of the cores and accretion disks of massive stars. The core size is $R_{\rm core} \simeq 0.06(M_{\rm core}/60M_\odot)^{1/2}\Sigma^{-1/2}$ pc (Note: $M_{60} = M_{\rm core}/60M_\odot$). Recall that $\Sigma$, the surface density of the clump, is related to the pressure of clump via $P \sim G\Sigma^2$. These small, pressure-confined cores have relatively small cross-sections for close interactions with other stars, although such interactions may still become important in the later stages of cluster formation once the stellar density has been built up to a high enough level. The rate of core collapse leading to accretion to the star, via a disk, is $\dot{m}_* = 4.6 \times 10^{-4} f_*^{1/2} M_{60}^{3/4} \Sigma^{3/4}\ M_\odot\ {\rm yr}^{-1}$, where $f_*$ is the ratio of $m_*$ to the final stellar mass and a 50% formation efficiency is assumed. Thus the collapse time, $1.3 \times 10^5 M_{60}^{1/4} \Sigma^{-3/4}$ yr, is short and quite insensitive to $M$, allowing coeval high and low mass star formation. The disk size is $R_{\rm disk} = 1200(\beta/0.02)(f_* M_{60})^{1/2}\Sigma^{-1/2}$AU, where $\beta$ is the initial ratio of rotational to gravitational energy of the core, and the normalization is taken from typical low-mass cores (Goodman et al. 1993), although there is quite a large dispersion about this value. These estimates allow quantitative models of the protostellar evolution, disk structure and outflow intensity. These have been compared to observations of the Orion KL protostar (Tan 2004a), also discussed briefly below. First I review other observational evidence of massive star formation from cores and accretion disks.

## Observational Evidence for Massive Star Formation from Cores and Accretion Disks

The issue of the mode of star formation, particularly massive star formation, is most likely to be resolved by observations. What observations are required? A common approach has been to search for disks around massive stars. However, these by themselves do not distinguish between the models, unless they are seen in conjunction with a collapsing pre-stellar core and it can be shown that the star accumulated most of its mass by accretion from the core through the disk. One would also like to show that the disk has maintained a fairly stable orientation, perhaps by looking at the impact of past outflow activity, during the accretion process, although even this is not necessarily to be expected from the collapse of a very turbulent core.

There are a number of claims for disks around massive protostars and massive young stars. Cesaroni et al. (1999) made mm and IR observations of IRAS 20126+4104, concluding the system showed the expected signatures of a massive ($\sim 24\ M_\odot$) protostar, forming from a Keplerian accretion disk inside a dense gas core. Shepherd, Claussen, & Kurtz (2001) used 7 mm observations to marginally resolve the driving source, G192.16, of a powerful molecular outflow, which from luminosity arguments is thought to be a $\sim 10\ M_\odot$ protostar. They interpreted the elongation, which is roughly perpendicular to the outflow, to be evidence for a $\sim$ 100 AU, $\sim 10\ M_\odot$ disk. However, much of the elongation is asymmetric, and so they also invoked a second protostar. Sandell, Wright, & Forster (2003) used 3.4 mm continuum and molecular line observations of NGC7538S to infer the presence of a rotating, massive ($\sim 100\ M_\odot$), and exceptionally large ($r_d \sim$ 14000 AU) disk about a $\sim 10^4\ L_\odot$ protostar, again driving a powerful outflow. This source is also peculiar in that, if it is a massive protostar, it is relatively isolated. Beltrán et al. (2004) used 1.4 mm continuum and molecular line observations to identify 4 massive protostellar disks by searching for velocity gradients perpendicular to outflows. They found disk sizes of several thousand AU. Chini et al. (2004) used NIR imaging and CO line observations in M17 to find an elongated structure $\sim 2 - 3 \times 10^4$ AU across with a mass of $\gtrsim 100\ M_\odot$ and a velocity gradient of $1.7\ \mathrm{km\ s^{-1}}$. In the above systems the velocities measured from molecular lines are typically on quite large scales that barely resolve the disk: the velocity differences are only a few $\mathrm{km\,s^{-1}}$, since the inner regions are not resolved. It is possible that some of these sources, particularly those where there is little evidence for a luminous central source or outflow, may simply be flattened or filamentary structures with a velocity gradient.

Pestalozzi et al. (2004) interpreted VLBI observations of methanol masers in NGC7538 IRS N1 in terms of an edge-on Keplerian disk extending to a radius of $\sim$ 1000 AU and orbiting a 30 $M_{\odot}$ protostar. While some methanol maser systems may trace accretion disks, it appears that many are in fact signatures of outflows (De Buizer 2003).

In more evolved and revealed systems, NIR spectra of CO and Br$\gamma$ emission have been used to infer the presence of disks, the emission coming from inside a few AU from the star (Blum et al. 2004; Bik & Thi 2004). Vink et al. (2002) used H$\alpha$ spectropolarimetry to show that Herbig Ae/Be stars are surrounded by flattened, presumably disk-like, structures. While studies of revealed systems are useful for probing the properties and lifetimes of remnant accretion disks, they do not directly test the different formation scenarios, since even stellar collisions would be expected to leave remnant material that would form a disk.

Most of the aforementioned systems are at distances of $\sim$ 2 kpc or more. The closest massive protostar is in the Orion KL region, only $\sim$ 450 pc away. Wright et al. (1996), Greenhill et al. (1998) and Tan (2004) have interpreted the system as containing a $r \sim$ 1000 AU accretion disk, as traced by SiO (v=0; J=2-1) maser emission, centered about the thermal radio source *I* (Menten & Reid 1995) and aligned perpendicular to the large scale molecular outflow that flows to the NW and SE. However, from SiO (v=1,2; J=1-0) masers within several tens of AU from source *I*, Greenhill et al. (2003) have interpreted the disk as being aligned parallel to the large scale outflow. In this case either the source is unrelated to the large scale outflow, or the orientation has changed in the last $\sim 10^3$ yr, the timescale of current outflow activity. Normally one would regard this last possibility as extremely unlikely, however, the motion of a $\sim$ 10 $M_{\odot}$ young star (the BN object) through the region occurred only 500 years ago (Plambeck et al. 1995; Tan 2004b). Several pieces of evidence point to an ejection of BN from the already-formed $\Theta^1 C$ binary system, however it is not possible to exclude an origin at source *I* itself (Bally & Zinnecker 2005; Rodriguez et al. 2005).

Outflows are common from regions of high-mass star formation (see Beuther & Shepherd 2005, these proceedings) for a review. However, because massive stars tend to be forming in clusters it is not always clear which sources are responsible for driving the outflows. Nevertheless there seems to be a multitude of collimated, powerful outflows, that appear to be scaled-up versions of those from low-mass protostars. The continuity in outflow properties from the low to high mass regimes suggests that there is a single driving mechanism (Beuther et al. 2002).

One difference between outflows from low-mass and high-mass protostars is the presence of high flux of ionizing radiation in the latter. This should create an "outflow-confined", hyper-compact HII region (Tan & McKee 2003). This model can account for the radio spectrum and morphology of source *I* in Orion KL, and perhaps also for the radio sources in CRL 2136 (Menten & van der Tak 2004), W33A, AFGL 2591 and NGC 7538 IRS9 (van der Tak & Menten 2005). An alternative model is the gravitationally-confined ionized accretion flow (Keto 2003), however this requires spherical accretion all the way to the star. Another model is the ionized flow from a photo-evaporated neutral disk (Hollenbach et al. 1994), however, if normal MHD outflows are present from the inner disk, they should block ionization of the outer disk.

## The timescale of star cluster formation

The timescales of star cluster formation have been reviewed by Tan (2005). Two independent pieces of evidence suggest that in the Orion Nebula Cluster, stars have been forming for at least 10 dynamical timescales, or 20 free-fall timescales. First ages of stars derived from pre-main-sequence tracks show a spread from 0 to at least 3 Myr (Palla & Stahler 1999). Second, the age of a dynamical ejection event of 4 massive stars ejected from a region coincident with the ONC is about 2.5 Myr (Hoogerwerf et al. 2001).

A relatively long formation timescale is also consistent with the observed morphologies of protoclusters in CS molecular lines: Shirley et al. (2003) find approximately spherical and centrally concentrated morphologies for a large fraction of their sources, suggesting they are older than a few dynamical times.

Formation timescales longer than a dynamical time allow the clump gas to virialize and come into pressure equilibrium: self-gravity is countered by internal sources of pressure. Numerical simulations (Mac Low et al. 1998; Stone et al. 1998) suggest that turbulence decays in one or two dynamical timescales (however, see Cho & Lazarian 2003). In this case, in order for turbulence support of the clump to be maintained, energy must be injected, most probably from internal sources such as protostellar outflows.

Such long formation timescales would also allow for significant dynamical relaxation of the forming star cluster: for $N$ equal mass stars the relaxation time is $t_{\rm relax} \simeq 0.1N/(\ln N)t_{\rm dyn}$, i.e. about 14 crossing timescales for $N = 1000$. Using numerical experiments, Bonnell & Davies (1998) found that the mass segregation time (of clusters with mass-independent initial velocity dispersions) was similar to the relaxation time. The presence of gas should shorten these timescales (Ostriker 1999). Therefore at least a part of the observed central concentration of massive stars in the Orion Nebula Cluster, in particular the Trapezium stars, may be due to mass segregation rather than preferential formation at the center.

It should be noted that a star cluster formation timescale of a few Myr is similar to the dynamical timescale of individual GMCs. Star formation appears to be rapid when compared to these timescales, but not when compared to the timescales of the star-forming clumps themselves. This is a major difference between the clustered (e.g. Orion) and distributed (e.g. Taurus, Hartman 2002) mode of star formation. Note also that even if the star cluster formation timescale is similar to the GMC dynamical timescale, this does not imply GMC lifetimes are this short (e.g. Tassis & Mouschovias 2004; §6).

## How does feedback affect the formation process?

Feedback processes that act against gravitational collapse and accretion of gas to protostars include radiation pressure (transmitted primarily via dust grains), thermal pressure of ionized regions and ram pressure from stellar winds, particularly MHD-driven outflows from protostars that are still actively accreting. If star cluster formation takes longer then $\sim 3$ Myr, then there is a chance of supernova feedback clearing out any remaining gas.

## Feedback in Individual Cores

For individual low-mass star formation from a core, bipolar protostellar outflows, accelerated from the inner accretion disk and star by rotating magnetic fields, appear to be the dominant feedback mechanism, probably preventing accretion from polar directions and also diverting a fraction, up to a third, of the material accreting through the disk. This leads to star formation efficiencies from the core of order 50% (Matzner & McKee 2000).

For massive protostars, forming in the same way from a core and accretion disk, one expects similar MHD-driven outflows to be present leading to similar formation efficiencies. In addition, once the massive protostar has contracted to the main sequence (this can occur rapidly before accretion has finished), it starts to produce large a flux of ionizing photons. The HII region is unlikely to be impeded by an accretion flow with reasonable angular momentum. However, it is likely to be confined, at least equatorially, by the bipolar outflow (Tan & McKee 2003). As the protostellar mass and ionizing flux increase, then eventually the HII region can spread through the outflow and start to ionize the disk surface. If the disk is ionized out to a radius where the escape speed is about equal to the ionized gas sound speed, then a photo-evaporated flow is set up, further reducing accretion to the star (Hollenbach et al. 1994).

Radiation pressure on dust grains (well-coupled to the gas at these densities) is also important for massive protostars. It has been suggested, in the context of spherical accretion models, that this leads to an upper limit to the initial mass function (Kahn 1974; Wolfire & Cassinelli 1987). However, these constraints are relaxed once a disk geometry is allowed for (Nakano 1989; Jijina & Adams

1996). Yorke & Sonnhalter (2002) used 2D axially symmetric simulations to follow massive star formation from a core collapsing to a disk, including radiation pressure feedback: accretion stopped at 43 $M_\odot$ in their most massive core. They showed the accretion geometry channeled radiative flux into the polar directions and away from the disk, terming this the "flashlight effect". Krumholz, McKee, & Klein (2005b) found that cavities created by protostellar outflows increase the flashlight effect, allowing even higher final masses.

## Feedback during Star Cluster Formation

One of the primary goals for models of feedback in star clusters is a prediction of the star formation efficiency, since this determines whether the cluster remains bound: Lada, Margulis, & Dearborn (1984) find from numerical models that clusters can remain bound with efficiencies as low as $\sim 30\%$ if the gas is removed gradually. Lada & Lada (2003) conclude that 90-95% of Galactic embedded clusters emerge from their GMCs unbound, although the mass associated with these systems is a somewhat lower percentage.

One can make some simple analytic estimates of the effects of massive star feedback on real protoclusters using Figure 1. The dashed line shows the condition that the escape speed at a distance $2R$ from the clump center is equal to the ionized gas sound speed ($\sim$ 10 $\mathrm{km\ s^{-1}}$). To the right and above this line ionizing feedback is much less effective since even if cluster gas is ionized it will be relatively difficult to be expelled. The Arches cluster in the Galactic center and super star clusters are in this region. If star clusters form relatively slowly, e.g. in $20t_{\rm ff}$ as may be the case in Orion (§5), then we can see that the clusters forming massive stars, which have approximately constant densities and free-fall timescales ($\sim 10^5$ yr), would not be affected by supernova feedback, since this only starts after at least $\sim 3 \times 10^6$ yr. However, the rate of star formation is uncertain, particularly in the more massive and distant clusters.

Matzner & McKee (2000) modeled protostellar outflow feedback in clusters of low-mass stars, estimating formation efficiencies of 30-50%. Adding high-mass stars to these particular models would presumably reduce the efficiency.

Tenorio-Tagle et al. (2003) presented a 1D model of star cluster formation in the presence of stellar wind and ionizing feedback, assuming an initial burst of massive star formation that creates a compressed shell from the infalling neutral gas, where more stars can form. They achieved high star formation efficiencies, allowing the build-up of very massive clusters, comparable to super star clusters. However, it is not clear how their model would fare in a more realistic turbulent and clumpy medium.

Scoville et al. (2001) considered radiation pressure feedback as a mechanism for limiting star cluster masses at $\sim 10^3\,M_\odot$. However, their model is 1D and it would be more difficult to disrupt gas if it were in optically thick clumps.

Tan & McKee (2001; 2004) used an idealized model to investigate feedback in a turbulent and clumpy medium. This structure was approximated by dividing the gas into cores and an intercore medium. The dynamics of the cores are affected by the potential of the overall protocluster and feedback effects from a stellar population at the cluster center, including radiation pressure, stellar winds and ionization, which can photo-evaporate cores. The main conclusion was that a clumpy, turbulent medium is much more capable of confining feedback, particularly ionizing feedback. HII regions are confined because they are continually injected with neutral cores that then suffer high photo-evaporation rates. This mass loading keeps the density and recombination rate relatively high. Such effects are likely to be important for Galactic ultra-compact ($\lesssim 0.1$ pc) HII regions, whose long lifetimes have been a puzzle. Figure 1 shows the parameters of three models, A, B, and C. Model A formed a cluster that dispersed its gas in 2 Myr, while B and C took about 3 Myr. Estimates of the star formation efficiency are somewhat uncertain as protostellar outflows were omitted. Within the limitations of the model, the efficiencies were $\sim 30\%$ for model A, and $\sim 50\%$ for models B and C.

## Feedback on GMCs

Once star clusters have formed, their feedback will impact the larger scale interstellar medium. In particular they may contribute to the destruction of GMCs. Williams & McKee (1997) considered the destruction of Galactic GMCs by photo-evaporation from ionizing photons from OB associations, finding destruction timescales of $\sim 30$ Myr for the most massive clouds. Matzner (2002) estimated slightly shorter destruction timescales for the same feedback process. Monaco (2004) considered the effects of supernova feedback on clouds that have already been shaped by ionization.

Alternatively, Ballesteros-Paredes (2004) argued that GMCs are transient phenomena and that their disruption is simply due to dynamical processes. Clark & Bonnell (2004) modeled star formation in such transient GMCs.

Observationally, Leisawitz, Bash & Thaddeus (1989) found that open star clusters older than about $\sim 10$ Myr were not associated with molecular clouds, which is consistent either with post-star-formation cloud lifetimes shorter than this age, i.e. only a couple of dynamical timescales, or with relative velocities of star clusters and their parent clouds of about $10\ \mathrm{km\,s^{-1}}$, which are to be expected from photoionization feedback (Williams & McKee 1997). The important question of GMC lifetimes remains open.

## Acknowledgments

I thank many colleagues for discussions, particularly Chris McKee and Mark Krumholz. I am supported by a Zwicky fellowship from ETH.

## References


Alves, J. F. Lada, C. J., & Lada, E. A. (2001). *Nature*, 409:159–161.

Ballesteros-Paredes, J. B. (2004). *Ap&SS*, 289:243–254.

Ballesteros-Paredes, J. B., Klessen, R. S., & Vázquez-Semadeni, E., (2003). *ApJ*, 592:188–202.

Bally, J., & Zinnecker, H. (2005). astro-ph/0502485.

Bate, M. R., Bonnell, I. A., & Price, N. M., (1995). *MNRAS*, 277:362–376.

Beltrán, M. T., Cesaroni, R., Neri, R., Codella, C., Furuya, R. S., Testi, L., & Olmi, L. (2004). *ApJ*, 601:L187–L190.

Bertoldi, F. and McKee, C. F. (1992). *ApJ*, 395:140–157.

Beuther, H., & Schilke, P. (2004). *Science*, 303, 1167

Beuther, H., & Shepherd, D. (2005). in *Cores to Clusters*, in press, (astro-ph/0502214).

Bik, A., & Thi, W. F. (2004) *A&A*, 427:L13–L16.

Blum, R. D., Barbosa, C. L., Damineli, A., Conti, P. S., & Ridgway, S. (2004). *ApJ*, 617:1167–1176.

Bonnell, I. A., Bate, M. R., & Zinnecker, H. (2004). *MNRAS*, 298:93–102.

Bonnell, I. A., Bate, M. R., Clarke, C. J., & Pringle, J. E. (2001). *MNRAS*, 323:785.

Bonnell, I. A., & Davies, M. B. (1998). *MNRAS*, 295:691.

Bonnell, I. A., Vine, S. G., and Bate, M. R. (2004). *MNRAS*, 349, 735.

Boulares, A., & Cox, D. P. (1990). *ApJ*, 365:544–558.

Carey, S. J., Clark, F. O., Egan, M. P., Price, S. D., Shipman, R. F., Kuchar, T. A. (1998). *ApJ*, 508, 721.

Cesaroni, R., Felli, M., Jenness, T., Neri, R., Olmi, L., Robberto, M., Testi, L., & Walmsley, C. M. (1999). *A&A*, 345:949–964.

Chini, R., Hoffmeister, V., Kimeswenger, S., Nielbock, M. Nürnberger, D., Schmidttobreick, L., & Sterzik, M. (2004). *Nature*, 429:155–157.

Cho, J., & Lazarian, A. (2003). *MNRAS*, 345:325–339.

Clark, P. C., & Bonnell, I. A. (2004). *MNRAS*, 347:L36–L40.

Crapsi, A., Caselli, P., Walmsley, C. M., Myers, P. C., Tafalla, M., Lee, C. W., & Bourke, T. L. (2005). *ApJ*, 619, 379–406.

Crutcher, R. M. (1999). *ApJ*, 520:706–713.

De Buizer, J. M. (2003). *MNRAS*, 341:277–298.

de Wit, W. J., Testi, L., Palla, F., & Zinnecker, H. (2005). *A&A*, in press, (astro-ph/0503337).

Deharveng, L., Zavagno, A., and Caplan, J. (2005). *A&A*, in press, astro-ph/0412602.

Dobbs, C. L., Bonnell, I. A., and Clark, P. C. (2005). *MNRAS*, in press.

Edgar, R., & Clarke, C. (2004). *MNRAS*, 349:678–686.

Egan, M. P., Shipman, R. F., Price, S. D., Carey, S. D., Clark, F. O., Cohen, M. (1998). *ApJ*, 494, L199.

Elmegreen, B. G. (2004). in *The Formation and Evolution of Massive Young Star Clusters*, ed. H. Lamers et al., ASP 322:277.

Elmegreen, B. G. and Lada, C. J. (1977). *ApJ*, 214:725–741.

Gilbert, A. M., & Graham, J. R. (2003). in Extragalactic Globular Clusters and their Host Galaxies, 25th IAU, JD 6.

Goodman, A. A., Benson, P. J., Fuller, G. A., & Myers, P. C. (1993). *ApJ*, 406:528.

Greenhill, L. J., Reid, M. J., Chandler, C. J., Diamond, P. J., & Elitzur, M. (2003). IAU Symp. 221, *Star Formation at High Angular Resolution*, eds. Burton et al., ASP, (astro-ph/0309334).

Hartmann, L. (2002). *ApJ*, 578:914–924.

Hollenbach, D. Johnstone, D., Lizano, S., & Shu, F. (1994). *ApJ*, 428:654.

Hoogerwerf, R., de Bruijne, J. H. J., & de Zeeuw, P. T. (2001). *A&A*, 365:49.

Jijina, J., & Adams, F. C. (1996). *ApJ*, 462:874.

Kahn, F. D. (1974). *A&A*, 37:149–162.

Keto, E. (2003). *ApJ*, 599:1196.

Kim, S. S., Figer, D. F., Lee, H. M., & Morris, M. (2000). *ApJ*, 545:301.

Klessen, R.. S., Ballesteros-Paredes, J. B., Vázquez-Semadeni, E., & Durán-Rojas, C. (2005). *ApJ*, 620:786–794.

Kroupa, P. (2002). *Science*, 295:82.

Krumholz, M. R., McKee, C. F., & Klein, R. I. (2005a). *ApJ*, 618:757–768.

Krumholz, M. R., McKee, C. F., & Klein, R. I. (2005b). *ApJ*, 618:L33–L36.

Lada, C. J. and Lada, E. A. (2003). *ARA&A*, 41:57–115.

Lada, C. J., Margulis, M., & Dearborn, D. (1984). *ApJ*, 285:141–152.

Larsen, S. S. (2002). *AJ*, 124:1393–1409

Larson, R. B. (2005). *MNRAS*, in press, astro-ph/0412357.

Leisawitz, D., Bash, F. N. & Thaddeus, P. (1989). *ApJS*, 70:731.

Li, D., Goldsmith, P. F., & Menten, K. (2003). *ApJ*, 587, 262

Li, P. S., Norman, M. L., Mac Low, M-M. and Heitsch, F. (2004). *ApJ*, 605:800–818.

Mac Low, M., & Klessen, R, S. (2004). *Reviews of Modern Physics*, 76:125–194.

Mac Low, M., Klessen, R, S., Burkert, A., & Smith, M. D. (1998). *Phys. Rev. Lett.*, 80:2754–2757.

Matzner, C. D. (2002) *ApJ*, 566:302–314.

Matzner, C. D., & McKee, C. F. (2000). *ApJ*, 545:364–378.

McKee, C. F. (1989). *ApJ*, 345:782–801.

McKee, C. F. (1999). *NATO ASIC Proc. 540: The Origin of Stars and Planetary Systems*, p29.

McKee, C. F. and Tan, J. C. (2002). *Nature*, 416:59–61.

McKee, C. F. and Tan, J. C. (2003). *ApJ*, 585:850–871.

Menten, K. M. & Reid, M. J. (1995). *ApJ*, 445:L157.

Menten, K. M. & van der Tak, F. F. S. (2004). *A&A*, 414:289–298.

Monaco, P. (2004). *MNRAS*, 354:151–160.

Motte, F., André, P., Ward-Thompson, D., & Bontemps, S. (2001). *A&A*, 372, L41

Mouschovias, T. Ch. (1996). in *Solar and Astrophysical Magnetohydrodynamic Flows*, ed. K. Tsiganos (Dordrecht: Kluwer), 505.

Mueller, K. E. and Shirley, Y. L. and Evans, N. J. and Jacobson, H. R. (2002). *ApJS*, 143:469–497.

Nakano, T. (1989). *ApJ*, 345:464–471.

Oey, M. S., Watson, A. M., Kern, K. and Walth, G. L. (2005). *AJ*, in press, astro-ph/0501136.

Palla, F., & Stahler, S. W. (1999). *ApJ*, 525:722.

Palous, J., Tenorio-Tagle, G. and Franco, J. (1994). *MNRAS*, 270:75.

Pestalozzi, M. R., Elitzur, M., Conway, J. E., & Booth, R. S. (2004). *ApJ*, 603:L113–L116.

Plambeck, R. L., Wright, M. C. H., Mundy, L. G., & Looney, L. W. (1995). *ApJ*, 455:L189.

Sandell, G., Wright, M., & Forster, J. R. (2003). *ApJ*, 590:L45–L48.

Scoville, N. Z., Polletta, M., Ewald, S., Stolovy, S. R., Thompson, R., & Rieke, M. (2001) *AJ*, 122:3017–3045.

Scoville, N. Z. and Sanders, D. B. and Clemens, D. P. (1986). *ApJ*, 310:L77–L81.

Shepherd, D. S., Claussen, M. J., & Kurtz, S. E. (2001). *Science*, 292:1513–1518.

Simon, R., Jackson, J. M., Clemens, D. P., Bania, T. M., & Heyer, M. H. (2001) *ApJ*, 551:747.

Solomon, P. M., Rivolo, A. R., Barrett, J., & Yahil, A. (1987). *ApJ*, 319:730–741.

Shirley, Y. L., Evans, N. J., Young, K. E., Knez, C. and Jaffe, D. T. (2003) *ApJS*, 149, 375

Shu, F. H., Adams, F. C., and Lizano, S. (1987). *ARA&A*, 25, 23.

Stone, J. M., Ostriker, E. C., & Gammie, C. F. (1998). *ApJ*, 508:L99.

Tan, J. C. (2000). *ApJ*, 536:173–184.

Tan, J. C. (2004a). in *Star Formation in the Interstellar Medium*, ASP, 323:249.

Tan, J. C. (2004b). *ApJ*, 607:47–50.

Tan, J. C. (2005). in *The Young Local Universe*, (astro-ph/0407093).

Tan, J. C., & McKee, C. F. (2001). in *Starburst Galaxies: Near and Far*, p188, (astro-ph/0012005).

Tan, J. C., & McKee, C. F. (2002). in *Hot Star Workshop III: The Earliest Phase of Massive Star Birth*, ASP 267, 267.

Tan, J. C., & McKee, C. F. (2003). IAU Symp. 221, *Star Formation at High Angular Resolution*, eds. Burton et al., ASP, (astro-ph/0309139).

Tassis, K. & Mouschovias, T. Ch. (2004). *ApJ*, 616:283–287.

Tenorio-Tagle, G., Palous, J., Silich, S., Medina-Tanco, G. A., & Muñoz-Tuñón, C. (2003) *A&A*, 411:397–404.

Testi, L., & Sargent, A. I. (1998). *ApJ*, 508, L91

Teyssier, D., Hennebelle, P., & Pérault, M. (2002). *A&A*, 382, 624.

Tremonti, C. A., Calzetti, D., Leitherer, C., & Heckman, T. M. (2001) *ApJ*, 555:322–337.

Turner, J. L., Beck, S. C., & Ho, P. T. P. (2000). *ApJ*, 532:L109.

van der Tak, F. F. S. & Menten, K. M. (2005). *A&A*, in press, astro-ph/0504026.

Vázquez-Semadeni, E., Kim, J., Shadmehri, M., & Ballesteros-Paredes, J. (2005). *ApJ*, 618, 344-359.

Walsh, A. J., Myers, P. C., & Burton, M. G. (2004). *ApJ*, 614, 194-202.

Whitworth, A. P. and Francis, N. (2002). *MNRAS*, 329:641–646.

Williams, J. P., & McKee, C. F. (1997). *ApJ*, 476:166–183.

Wolfire, M. G., & Cassinelli, J. P. (1987). *ApJ*, 319:850–867.

Wright, M. C. H., Plambeck, R. L., Mundy, L. G., & Looney, L. W. (1995). *ApJ*, 455:L185.

Yorke, H. W., & Sonnhalter, C. (2002). *ApJ*, 569:846–862.

RayJay and Henrik Beuther. Hugo Ledo and Monika Petr-Gotzens at the back

# PRECURSORS OF UCHII REGIONS AND THE EVOLUTION OF MASSIVE OUTFLOWS

Henrik Beuther
*Harvard-Smithsonian Center for Astrophysics*
*60 Garden Street*
*Cambridge, MA 02138*
*USA*
hbeuther@cfa.harvard.edu

Debra Shepherd
*National Radio Astronomy Observatory*
*Socorro, NM 87801*
*USA*
dshepher@nrao.edu

**Abstract** This paper covers two subjects in the field of massive star formation which in its details can be discussed separately. Therefore, we present first characteristics of precursors of UCHII regions and their likely evolutionary properties. The second section discusses massive molecular outflows, their implications for high-mass star formation, and a possible evolutionary sequence for massive outflows.

## Precursors of UCHII regions

Evolutionary scenarios suggest that High-Mass Starless Cores (HMSCs) represent the earliest evolutionary stage of massive star formation (e.g., Evans et al., 2002). The next observable evolutionary stage is when a High-Mass Protostellar Object (HMPO) forms in the core producing strong millimeter continuum and mid-infrared emission but no detectable centimeter continuum due to free-free emission from ionized gas (e.g., Molinari et al., 1996; Sridharan et al., 2002). The lack of detectable free-free emission may be because the protostar has not reached the main sequence or because the accretion rate onto the stellar surface is high enough to quench a developing HII region (e.g., Walmsley, 1995; Churchwell, 2002). Hot Cores are considered to belong to the HMPO stage. Soon after the hot core is formed, the central massive protostar produces adequate ionizing radiation to form a hypercompact, unresolved, most likely optically thick HII region (HCHII). The HCHII region may still be partially quenched by infalling gas which can hinder the expansion of the hypercompact HII region (e.g., Walmsley, 1995; 2003). Eventually, the hypercompact HII region begins to expand forming the well-studied ultracompact HII region (e.g., 1989; Kurtz et al., 1994) and later a more evolved HII region. Theoretical model calculations estimate an evolutionary time-scale for massive star formation of about $10^5$ yrs, assuming that massive stars form via accretion-based processes similar to low-mass star formation (McKee and Tan, 2002). A key difference between low- and high-mass star formation is that massive stars reach the main sequence while still actively accreting; low-mass stars end their accretion phase before they reach the main sequence. Here, we discuss the properties of the precursors of ultracompact HII regions, namely HMPOs and HMSCs.

## High-Mass Protostellar Objects (HMPOs)

There are several review articles that discuss the earliest stages of massive star formation and cover different aspects of the topic in more depth than possible here (e.g., Garay and Lizano, 1999; Kylafis and Pavlakis, 1999; Kurtz et al., 2000; Churchwell, 2002; Cesaroni, 2004). Therefore, we focus on the basic characteristics of HMPOs and discuss a few recent results in this field.

**Basic characteristics.** The far-infrared colors of HMPOs are similar to those of UCHII regions (e.g., 1989; Sridharan et al., 2002). This is mainly due to the fact that they resemble each other with regard to their gas/dust temperatures and the presence of dense gas. A major observational difference is that HMPOs show no or only weak cm continuum emission, thus the central objects have not yet produced a significant ionized region to trigger the necessary free-free emission. While most of the luminosity from UCHII regions is due to a central H-burning star, a large fraction of the luminosity from HMPOs is still expected to be due to accretion (e.g., Sridharan et al., 2002). Furthermore, the average observed $NH_3(1,1)$ linewidth toward HMPOs is $\Delta v \sim 2.1$ km/s compared to average values toward UCHII regions of $\Delta v \sim 3.1$ (Sridharan et al., 2002). This difference implies less turbulent motions at the HMPO stage.

Furthermore, $H_2O$ and/or Class II $CH_3OH$ maser emission is frequently observed toward HMPOs (e.g., Walsh et al., 1998; Beuther et al., 2002d; Codella et al., 2004). The existence of one or the other type of maser emission is widely regarded as a good signpost for massive star formation (in spite of $H_2O$ maser emission being also observed toward low-mass star-forming regions), but nevertheless, there is no general agreement about the physical processes generating the maser emission. There is indicative evidence that both maser types are produced either within accretion disks or molecular outflows (e.g., Norris et al., 1998; Walsh et al., 1998; Torrelles et al., 1998; Minier et al., 2000; De Buizer et al., 2000; De Buizer, 2003; Codella et al., 2004). Although $H_2O$ and Class II $CH_3OH$ masers are produced during the early stages of massive star formation, it appears that in general they cannot be used to derive the underlying physical processes of the regions. This statement does not exclude the possibility that in selected sources very high-spatial resolution and proper motion studies of masers can constrain the processes triggering the maser emission (for prominent recent examples see, e.g., Torrelles et al., 2003; Pestalozzi et al., 2004).

**Millimeter continuum emission.** Observations of HMPOs in (sub)mm continuum emission have shown that they exhibit strong dust continuum emission from massive dust and gas cores (e.g., Molinari et al., 2000; Beuther et al., 2002b; Mueller et al., 2002; Williams et al., 2004). Even with the rather coarse spatial resolution of most single-dish investigations ($\geq 11''$ beam), the gas cores often show filamentary and multiple structures. The density distribution of the cores are usually well fitted by power-law distributions and the power-law indices are in most cases around 1.6 (e.g., Beuther et al., 2002b; Mueller et al., 2002), similar to the power-law indices of young low-mass cores (e.g., Motte et al., 1998). Furthermore, multi-wavelength investigations show that the dust opacity index $\beta$ toward HMPOs is significantly lower than toward UCHII regions (mean values of 0.9 versus 2.0, Williams et al., 2004; Hunter, 1997). In low-mass cores, low dust opacity indices are attributed either to increasing opacity or grain growth at early evolutionary stages (e.g., Hogerheijde and Sandell, 2000; Beckwith et al., 2000). In contrast, Williams et al. (2004) show for their sample that the lowest observed values of $\beta$ are associated with lower rather than higher opacities. Therefore, these observations are consistent with grain growth in the early evolutionary stages of the dense cores. However, a point of caution remains because significant grain growth is expected to take at least $10^5$ years (e.g., Ossenkopf and Henning, 1994) whereas the whole massive star formation process might only last for $\sim 10^5$ years (McKee and Tan, 2002). If these time-scales are correct it may be difficult to produce similarly large grains in high-mass cores as found toward their low-mass counterparts, and the observed low values of $\beta$ might require a different interpretation.

**Fragmentation and the Initial Mass Function (IMF).** One important question for the whole cluster formation process is: At what evolutionary stage does the Initial Mass Function gets established? Do the initial fragmentation processes of the massive cores already show signatures of the IMF, or is competitive accretion during the ongoing cluster formation the main driver of the IMF (e.g., Bonnell et al., 2004)? In the last two years, a few different studies have been made to investigate the protocluster mass distribution and its connection to the IMF: Shirley et al., 2003, analyze single-dish CS observations toward a sample of 63 massive $H_2O$ maser sources with $\sim 30''$ resolution, Williams et al., 2004, do a similar study based on single-dish (sub)mm continuum observations with $8'' - 15''$ resolution, and Beuther and Schilke, 2004, observe one HMPO in the mm continuum emission at $1.5'' \times 1''$ resolution with the Plateau de Bure Interferometer. The main difference between the two

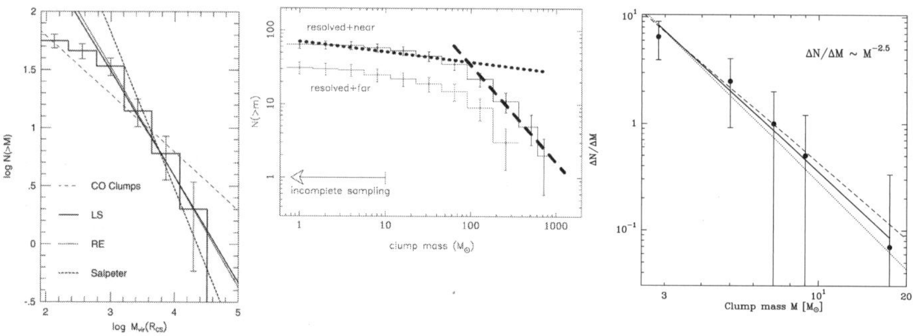


*Figure 1.* Compilation of mass distributions from very young high-mass star-forming regions: the cumulative mass distribution to the left is from single-dish CS observations of a sample of massive $H_2O$ maser sources (Shirley et al., 2003, LS and RE correspond to Least Square and Robust estimation fits). The cumulative mass spectrum in the middle is from single-dish mm dust continuum observations of a sample of HMPOs (Williams et al., 2004, the dotted and full lines correspond to far and near kinematic distances, respectively). The protocluster mass function to the right is from high-spatial resolution PdBI observations of one HMPO IRAS 19410+2336 (Beuther and Schilke, 2004, the full line is the best fit to the data, and the dashed and dotted lines present the IMFs derived from Salpeter, 1955, and Scalo, 1998).

single-dish studies and the interferometric investigation is that the single-dish observations average out whole young clusters and thus need many clusters to derive a cumulative mass distribution of the sample of protoclusters, whereas Beuther & Schilke (2004) only investigate one young cluster but are capable of resolving many sources within this cluster. All of the above mentioned studies preferentially select regions that are in the earliest evolutionary stages of star formation and most have not yet formed detectable UCHII regions. Figure 1 summarizes the results from the three studies.

Despite the intrinsic differences in completeness limits between the Shirley et al. and Williams et al. studies ($> 1000$ and $> 10\,M_\odot$, respectively), both studies show power-law distributions comparable to the stellar IMF. On much smaller spatial scales, Beuther & Schilke (2004) also find a good correspondence between the protocluster mass spectrum in IRAS 19410+2336 compared with the IMFs as derived by, e.g, Salpeter, 1955; Scalo, 1998. These independently derived results are encouraging support for the idea that the early fragmentation of massive star-forming cores is responsible for the determination of the IMF, and not processes which might take place at evolutionary later stages like competitive accretion. An additional result from the single-dish observations alone is that the star-formation efficiency for the sample of protoclusters is relatively mass invariant (Williams et al., 2004).

**Chemistry.** Young massive star-forming regions and especially Hot Cores are known to have an incredibly rich chemistry (e.g., the rich spectra toward the Orion-KL Hot Core, e.g., Blake et al., 1987; Schilke et al., 1997). Here, we present recent submillimeter observations of the prototypical Hot Core

Orion-KL. A more detailed discussion of core chemistry is presented in Paola Caselli's contribution to this volume.

Beuther et al., 2004b, used the Submillimeter Array (SMA) to image the submm line and continuum emission of Orion-KL at sub-arcsecond resolution. In the submm continuum, the enigmatic source "I" could be disentangled from the Hot Core, and source "n" was detected for the first time shortward of 7 mm (Beuther et al., 2004b). Additionally, within one spectral setup of the SMA (4 GHz combining the upper and lower sideband) approximately 150 molecular lines were observed (Beuther et al. in prep.). Figure 2 shows example line images which highlight the spatial variations observed between the different molecular species: SiO is detected mainly from the outflow emanating from source "I", but all other molecules peak offset from source "I". Typical hot

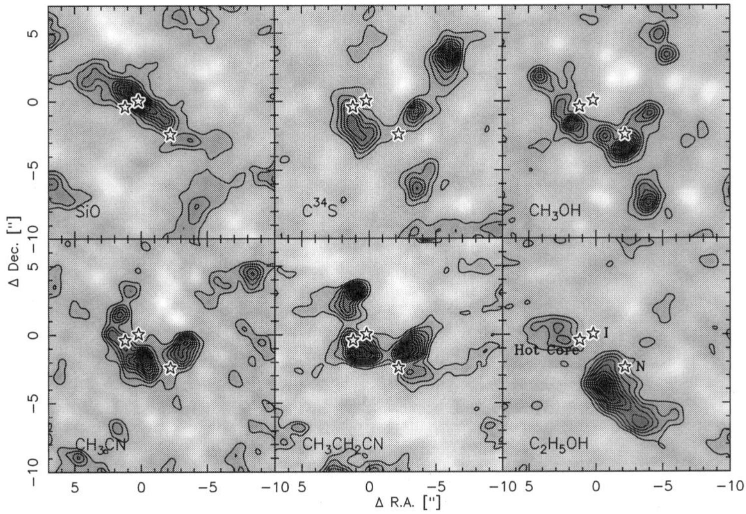


*Figure 2.* Submm line images obtained with the SMA at 865 $\mu$m (Beuther et al. in prep.). The grey-scale and contours show the molecular emission as labeled at the bottom left of each panel. The three stars mark the positions of source "I", the Hot Core mm continuum peak, and source "n" (labeled at the bottom-right).

core molecules like $CH_3CN$ or $CH_3CH_2CN$ show the hot core morphology previously reported from, e.g., $NH_3$ observations (Wilson et al., 2000). Contrary to this, $CH_3OH$ does show some hot core emission as well, but there is an additional strong $CH_3OH$ peak further south associated with the so called compact ridge emission. Other molecules like $C^{34}S$ or $C_2H_5OH$ show even different spatial emission, tracing partly the hot core and the compact ridge but also showing an additional peak to the north-west that is spatially associated with IRC6. We still do not properly understand the underlying reasons for this spatial molecular diversity, but these observations of the closest Hot

Core, Orion-KL, stress that we must be cautious in our interpretations of spatial structures for other Hot Cores that are typically at much greater distances.

## High-Mass Starless Cores (HMSCs)

Prior to the infrared satellites ISO and MSX, there was no observational evidence for High-Mass Starless Cores, and HMPOs were considered to be the youngest observationally accessible stage of massive star formation. The detection of Infrared Dark clouds (IRdCs) and their association with mm dust continuum emission changed that situation (e.g., Egan et al., 1998), and since then observational studies of these HMSCs/IRdCs have begun to shed first light on this evolutionary stage. Evans et al., 2002, outlined potential properties of HMSC, and recently Garay et al., 2004, presented an analysis of four such sources. Here, we present an approach to identify and study HMSCs based on a comparison of previously derived mm continuum maps of HMPOs (Beuther et al., 2002b) with the MSX database (Sridharan et al. in prep.).

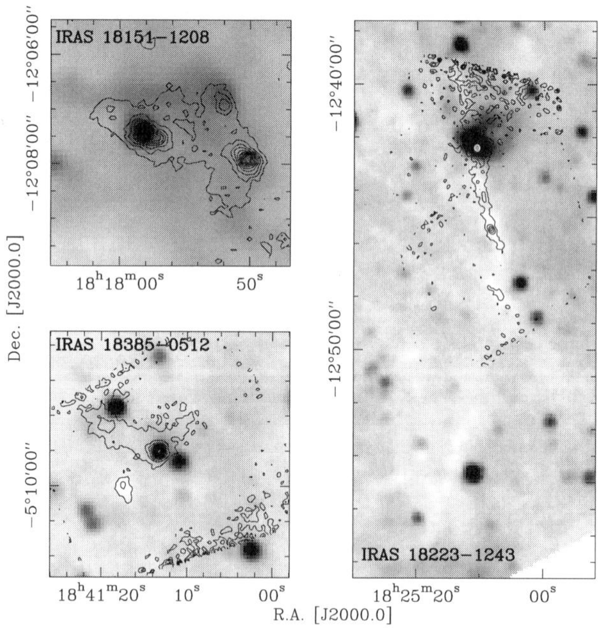


*Figure 3.* The grey-scale shows the 8 $\mu$m emission from the MSX satellite, and the contours present the 1.2 mm continuum emission first presented in Beuther et al., 2002b.

Based on an initial study of HMPOs associated with IRAS sources (Sridharan et al., 2002), Beuther et al., 2002b, observed serendipitously within the same fields of view a number of secondary mm-peaks which were not associated with IRAS sources. Correlation of these mm-peaks with the MSX mid-infrared observations revealed a number of mm-peaks which are not only weak in the mid-infrared but which are seen as absorption shadows against the galactic background (Fig. 3, Sridharan et al. in prep.).

Observing these HMSCs with the Effelsberg 100 m telescope in $NH_3$ (Sridharan et al. in prep.), the derived average temperatures are around 15 K compared to the average temperatures of the more evolved HMPOs of 22 K. Furthermore, we find on average a smaller linewidth of 1.6 km/s toward the HMSCs in comparison to a mean value of 2.1 km/s observed toward the HMPOs. With gas masses of the order a few hundred $M_\odot$, no mid-infrared emission, cold temperatures and less turbulence implied by the small line-widths, these sources are promising candidates for being HMSCs on the verge of forming massive clusters. Ongoing single-dish and interferometric follow-up observations of this and other samples will hopefully improve our understanding of the initial conditions of massive star formation significantly.

## Massive molecular outflows: an evolutionary scenario

In contrast to low-mass outflows, there is no general census for the understanding of massive molecular outflows. For recent reviews on different aspects of massive molecular outflows see Bachiller and Tafalla, 1999; Churchwell, 1999; Richer et al., 2000; Shu et al., 2000; Konigl and Pudritz, 2000; Shepherd, 2003; Beuther, 2004.

Two competing scenarios have been put forth to explain the formation of massive stars: one possibility is to extend the low-mass models to high-mass star formation and form massive stars via enhanced accretion processes through accretion disks (e.g., Jijina and Adams, 1996; Norberg and Maeder, 2000; Yorke and Sonnhalter, 2002; McKee and Tan, 2003). Contrary to this, at the dense centers of evolving massive clusters, competitive accretion could become dominant and it might even be possible that (proto)stellar mergers occur and form the most massive objects via coagulation processes (e.g., Bonnell et al., 1998; Stahler et al., 2000; Bally, 2002; Bonnell et al., 2004). Massive molecular outflows can help to differentiate between these scenarios: while the enhanced accretion model predicts collimated molecular outflows with properties similar to those for low-mass outflows, it is unlikely that highly-collimated structures could exist in the coalescence scenario and the outflow energetics would be significantly different.

## Summary of observational constraints

Observations of massive molecular outflows are unfortunately not conclusive. On the one hand, some observations find that molecular outflows from high-mass star-forming regions appear less collimated than outflows from low-mass regions (e.g., Shepherd and Churchwell, 1996a; Ridge and Moore, 2001;

Wu et al., 2004). However, it has been shown that the lower collimation factors of massive outflows observed with single-dish telescopes could simply be due to the on average larger distances of high-mass star-forming regions (Beuther et al., 2002c). A few interferometric observations with $0.1'' - 10''$ resolution suggest that outflow opening angles from at least some young, early-B stars can be significantly wider than those seen toward low-mass protostars (e.g., Shepherd et al., 1998; Shepherd et al., 2001; Shepherd et al., 2003). Thus, greater distances and the associated lower angular resolution is not the only reason for the observed poor collimation in some massive outflows. There is also evidence that the energetics in at least one massive outflow from an early B star in W75N are different to those in low-mass outflows, although exactly what this implies is unclear (Shepherd et al., 2003).

In recent years a few examples of collimated massive outflows from early B stars have been found (e.g., Beuther et al., 2002a; Gibb et al., 2003; Su et al., 2004). A comparison of position-velocity diagrams from collimated massive molecular outflows with their low-mass counterparts reveal similar signatures from outflows of all masses (Beuther et al., 2004a). Moreover, a near-infrared analysis of a HMPO found collimated outflows with properties appearing to be scaled-up versions of low-mass outflows (Davis et al., 2004). This latter group of observations favors an accretion scenario for the formation of massive stars that is a scaled up version of the low-mass scenario.

For both well-collimated and poorly collimated outflows, relations between outflow and source parameters hold over many orders of magnitude (e.g., Richer et al., 2000; Beuther et al., 2002c; Wu et al., 2004). Richer et al. (2000) point out that these correlations do not necessarily argue for a common entrainment or driving mechanism for sources of all luminosities given that one would expect that most physically reasonable outflow mechanisms would generate more powerful winds if the source mass and luminosity are increased. It is interesting to note that no well-collimated outflows have been observed toward a (proto)-O star (e.g., Shepherd, 2003, and references therein, Sollins et al., 2004). Thus, the link between accretion and outflow is not as well established for forming O stars as for early B stars and formation by coalescence may still be a possibility. Table 1 presents a number of relevant references for observations which shed light on massive outflows.

The current situation is that different observations find results which appear to contradict each other. In order to resolve this conflict, it is necessary to re-evaluate the observations. In principle, three possibilities exist to solve the discrepancies: (a) some observations are wrongly interpreted, (b) there exist physically different modes of massive molecular outflows and thus massive star formation, or (c) the observed sources are not directly comparable, e.g., there might be evolutionary differences between the observed sources which could translate into various evolutionary signatures of the massive outflows.

*Table 1.* Summary of outflow results. This table does not claim completeness but gives an overview of relevant publications connected to the question of outflow evolution.

| Source | Results | Ref |
|---|---|---|
| | Low-spatial resolution single-dish studies | |
| 122 UCHIIs | ubiquitous phenomena | 1 |
| 10 UCHIIs | 50% bipolarity, low collimation, $\dot{M}$ vs $L_{\rm bol}$ relation | 2 |
| 3 YSOs | $\dot{M}$ vs $L_{\rm bol}$ still holds | 3 |
| 11 YSOs | low collimation, no Hubble law, $\dot{M}$ vs $L_{\rm bol}$ not correlated | 4 |
| 69 HMPOs | 55% bipolarity, accretion-outflow process suggested | 5 |
| 26 HMPOs | 80% bipolarity, no $\dot{M}$ vs $L_{\rm bol}$ but $M_{\rm out}$ vs $M_{\rm core}$, consistent with collimated outflows, similar to low-mass outflows | 6 |
| 139, all types | $M_{\rm out}$ & $\dot{M}$ correlate with $L_{\rm bol}$, on average less collimated | 7, 8 |
| | High-spatial resolution interferometer and infrared studies | |
| G192.16, HCHII | wide-angle wind + molecular flow & disk | 9,10,11 |
| IRAS20126, HCHII | precessing jet with disk | 12,13 |
| IRAS05358, HMPO | consistent with low-mass jets | 14 |
| W75N, HC/UCHII | energetics different to low-mass, not scaled up outflows | 15 |
| IRAS19410, HMPO | consistent with low-mass outflows | 16 |
| G35.2, HCHII | a cluster of collimated outflows | 17 |
| 2 HMPOs | accretion based outflows | 18 |
| 4 HMPOs/HCHIIs | same physical driving mechanisms as low-mass outflows | 19 |
| IRAS16547, HMPO | collimated jet, similar to low-mass jets | 20,21 |
| IRAS18151, HMPO | scaled-up versions of low-mass jets driven by disk accretion | 22 |
| G5.89, UCHII | multiple wide-angle outflows, harbors O5 star | 23 |

The listed results just refer to mm studies of thermal lines and infrared studies. Ref: (1) Shepherd and Churchwell, 1996b, (2) Shepherd and Churchwell, 1996a, (3) Henning et al., 2000, (4) Ridge and Moore, 2001, (5) Zhang et al., 2001, (6) Beuther et al., 2002c, (7) Wu et al., 2004, (8) Wu et al., 2005, (9) Shepherd et al., 1998, (10) Devine et al., 1999, (11) Shepherd et al., 2001, (12) Cesaroni et al., 1999, (13) Shepherd et al., 2000, (14) Beuther et al., 2002a, (15) Shepherd et al., 2003, (16) Beuther et al., 2003, (17) Gibb et al., 2003, (18) Su et al., 2004, (19) Beuther et al., 2004a, (20) Garay et al., 2003, (21) Brooks et al., 2003, (22) Davis et al., 2004, (23) Sollins et al., 2004

It is possible that (a) applies to some of the single-dish results, e.g., the initial claim of low collimation for massive molecular outflows can be misinterpreted by the low spatial resolution of the observation, the generally larger distances to the sources and uncertainties due to the unknown inclination angles of the outflows. However, the results based on high-spatial-resolution observations are more significant and cannot be discarded in this manner. It appears that physically different modes and/or evolutionary differences must exist to explain all the observations. While we cannot exclude the possibility that different modes of star formation and their associated outflows exist, it seems unlikely given the link between accretion and outflow that has been established up to early B spectral types. Thus, here we propose an evolutionary scenario for massive outflows which is capable of explaining the morphological and energetic results within a consistent picture of massive protostellar evolution. This evolutionary scenario is based on the idea that stars of all masses form via similar accretion-based processes that are proposed for low-mass star formation (e.g., McKee and Tan, 2003).

## A potential evolutionary scenario

Re-evaluation of the high-spatial resolution literature presented in Table 1 shows that most of the youngest early B protostars (HMPOs and those producing hypercompact HII regions) support the idea that these high-mass outflows are driven by similar physical processes as those seen in low-mass outflows. Contrary to this, three sources in particular show different observational characteristics: G5.89: which harbors an O5 star and multiple wide-angle outflows, and G192.16 and W75N: two early B stars with UCHII regions. G192.16 and W75N VLA2 both appear to be relatively old (a few $\times 10^5$ years based on the CO dynamical timescale), and they have had adequate time to reach the main sequence and produce significant ionizing radiation. The outflows are poorly collimated within 100 AU of the central star and their outflow morphology on larger scales are consistent with wind-blown bubbles. Furthermore, it should be noted that the single-dish outflow studies, which report a lower degree of collimation for massive outflows preferentially observed sources with UCHII regions (Shepherd and Churchwell, 1996a; Ridge and Moore, 2001). In contrast, most other single-dish surveys were directed toward samples of younger sources, namely HMPOs, observing higher collimation degrees consistent with low-mass outflows.

To date, the different observations have been interpreted either in favor of: (1) similar driving processes for outflows of all masses (e.g., Beuther et al., 2002c; Davis et al., 2004); or (2) that massive stars form via different physical processes like explosive coalescence or that the outflows are powered by a combination of accretion and deflected outflow. (e.g., Churchwell, 1999; Bally, 2002). Here, we propose a different, less controversial interpretation of the data: an evolutionary sequence for massive protostars. We consider two possible evolutionary sequences which could result in similar observable outflow signatures as shown in Figure 4.

*(a) The outflow evolution of an early B-type star:* In the accretion scenario, a B star forms via accretion through a disk. During its earliest HMPO phase no HCHII region has formed yet, and the disk outflow interaction produces collimated jet-like outflows (Fig. 4 (left), e.g., IRAS05358). At some point, a HCHII forms and the wind from the central massive star produces an additional less collimated outflow component. At that stage the disk is not entirely destroyed yet, and jet and wind can co-exist (Fig. 4 (center), e.g., IRAS 20126). Evolving further, a typical ultracompact HII region forms above and below the accretion disk and the massive wind begins to dominate the whole system (Fig.

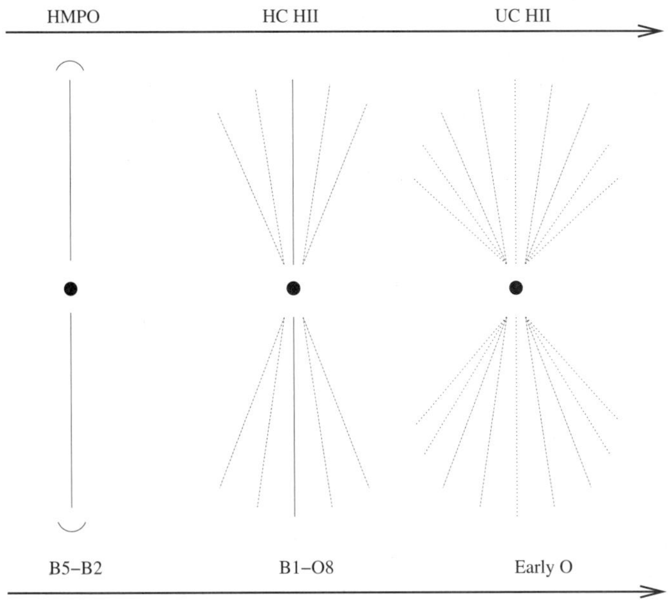


*Figure 4.* Sketch of the proposed evolutionary outflow scenario. The three outflow morphologies can be caused by two evolutionary sequences: (top) the evolution of an early B-type star from an HMPO via a HCHII region to an UCHII region, and (bottom) the evolution of an early O-type star which goes through B- and late O-type stages (only approximate labels) before reaching its final mass and stellar luminosity.

4 (right), e.g., W75N). Remnants of the initial jet might still exist, and density gradients within the maternal core are likely to maintain at least a small degree of collimation at that stage.

*(b) The outflow evolution of an early O-type star:* An intriguing aspect of the outflow studies is that so far extremely collimated jet-like outflows have not been observed toward sources earlier than B1. It is possible that this is simply a selection effect because the evolutionary timescale of early O stars is expected to be only around $10^5$ years (McKee and Tan, 2002). Given the scarcity of O stars and the short formation time, it is difficult to find sources while they are still in the earliest stages of formation. However, to form massive stars via accretion, the protostellar objects must accrete even after the central object has reached the main sequence. In this scenario, at the beginning, the massive protostar reaches a mid-B protostar-like state where it drives collimated outflows (Fig. 4 (left)). As the object continues to accrete it becomes an early-B to late-O star, forming a HCHII region while the wind and jet components could co-exist (Fig. 4 (center)). Finally, as the central star evolves, it may eventually become a mid- to early-O star. The increased radiation from the central star would generate significant Lyman-$\alpha$ photons and ionize the molecular outflowing gas even at large radii. The result may be an increase in the

plasma pressure at the base of the outflow which could overwhelm the collimating effects of a magnetic field (e.g., Shepherd, 2003; Konigl and Pudritz, 2000). Incorporating the additional physical effects of high-mass protostars on the surrounding disk and envelope is one of the challenges of theoretical massive outflow modeling (e.g., Konigl, 1999). In this scenario, it would be intrinsically impossible to ever observe collimated jet-like outflows from very young early O-type (proto)stars.

## Discussion and potential caveats

This scenario is not the only possible interpretation, and different physical processes could take place at the earliest stages of massive star formation. This re-evaluation is based only on imaging data of thermal lines in the mm and infrared bands. However, many studies of outflows have also been undertaken by the means of maser studies, especially $H_2O$ maser. While some maser studies found $H_2O$ maser emission consistent with jet-like outflows (e.g. Kylafis and Pavlakis, 1999), other studies find shell-like expanding features which cannot be explained by collimated jets (e.g., Patel et al., 2000; Torrelles et al., 2001; Torrelles et al., 2003). The most intriguing case is the $H_2O$ maser study in W75N: Torrelles et al. (2003) find on projected spatial scales of 1400 AU that the two cm continuum sources VLA1 and VLA2 drive very different types of outflows, the first one appears to be a collimated jet whereas the second one rather resembles a wind-like shell. As the two sources are part of the same core, environmental properties cannot explain the difference. Both sources appear to have a similar luminosity (B1-B2 spectral types) although the spectral type derived for VLA1 should be considered an upper limit due to likely strong contamination by ionizing flux produced by shock waves in the jet (Shepherd et al., 2004). Assuming the VLA1 jet is produced by an early B star, then the difference in morphology would not be a question of the mass of the central source as well. Thus, Torrelles et al. (2003) propose that an evolutionary explanation is most likely. They cannot determine which source is younger but based on the extent of the maser emission – the non-collimated features are more compact – they suggest that the collimated jet might belong to the more evolved source. This would be a counter-example to the scenario presented in this paper as well as to the scenarios discussed for low-mass outflows (e.g., Andre et al., 2000). However, maser emission appears to be very selective, and many known outflow sources have no $H_2O$ maser emission at all (e.g., Beuther et al., 2002c). Therefore, it is also possible that the collimated part of the outflow from VLA2 is simply not depicted by the data, and the non-collimated features have been excited more recently by the central HCHII region. Furthermore, the cm continuum emission from VLA1 may be due primarily to the jet (Shepherd et al., 2004). Thus, it is feasible that the central source of VLA1 is either less massive than VLA2 or it is younger and has not formed a HCHII region yet. Therefore, we might only observe a collimated feature toward this source. The latter scenario is consistent with the evolutionary sequence presented here. To properly differentiate between both scenarios in this source requires that we find other means to determine the age of the powering sources VLA1 and VLA2.

It should be stressed that the proposed evolutionary scenario is still only a potential, qualitative explanation for the observed outflow features which must be tested against both theory and observations.

## References


Andre, P., Ward-Thompson, D., and Barsony, M. (2000). *Protostars and Planets IV*, pages 59.

Bachiller, R. and Tafalla, M. (1999). In *NATO ASIC Proc. 540: The Origin of Stars and Planetary Systems*, pages 227.

Bally, J. (2002). In *Hot Star Workshop III: The Earliest Stages of Massive Star Birth. ASP Conference Proceedings, Vol. 267. Edited by Paul A. Crowther.*, pages 219.

Beckwith, S. V. W., Henning, T., and Nakagawa, Y. (2000). *Protostars and Planets IV*, pages 533.

Beuther, H. (2004). *in: X-Ray and Radio Connections, astro-ph/0404563.*

Beuther, H. and Schilke, P. (2004). *Science*, 303:1167–1169.

Beuther, H., Schilke, P., and Gueth, F. (2004a). *ApJ*, 608:330–340.

Beuther, H., Schilke, P., Gueth, F., et al. (2002a). *A&A*, 387:931–943.

Beuther, H., Schilke, P., Menten, K. M., et al., (2002b). *ApJ*, 566:945–965.

Beuther, H., Schilke, P., Sridharan, T. K., et al., (2002c). *A&A*, 383:892–904.

Beuther, H., Schilke, P., and Stanke, T. (2003). *A&A*, 408:601–610.

Beuther, H., Walsh, A., Schilke, P., et al., (2002d). *A&A*, 390:289–298.

Beuther, H., Zhang, Q., Greenhill, L. J., et al., (2004b). *ApJ*, 616:L31–L34.

Blake, G. A., Sutton, E. C., Masson, C. R., and Phillips, T. G. (1987) *ApJ*, 315:621.

Bonnell, I. A., Bate, M. R., and Zinnecker, H. (1998). *MNRAS*, 298:93–102.

Bonnell, I. A., Vine, S. G., and Bate, M. R. (2004). *MNRAS*, 349:735–741.

Brooks, K. J., Garay, G., Mardones, D., and Bronfman, L. (2003). *ApJ*, 594:L131–L134.

Cesaroni, R., Felli, M., Jenness, T., Neri, R., Olmi, L., Robberto, M., Testi, L., Walmsley, C. M. (1999) *A&A*, 345:949-964.

Cesaroni, R. (2004). *in: Astrophysics and Space Science, in press.*

Churchwell, E. (1999). In *NATO ASIC Proc. 540: The Origin of Stars and Planetary Systems*, pages 515.

Churchwell, E. (2002). *ARA&A*, 40:27–62.

Codella, C., Lorenzani, A., Gallego, A. T., Cesaroni, R., and Moscadelli, L. (2004). *A&A*, 417:615–624.

Davis, C. J., Varricatt, W. P., Todd, S. P., and Ramsay Howat, S. K. (2004). *A&A*, 425:981–995.

De Buizer, J. M., Piña, R. K., Telesco, C. M. (2000). *ApJS*, 130:437–461.

De Buizer, J. M. (2003). *MNRAS*, 341:277-298.

Devine, D., Bally, J., Reipurth, B., et al., (1999). *AJ*, 117:2919–2930.

Egan, M. P., Shipman, R. F., Price, S. D., et al., (1998). *ApJ*, 494:L199+.

Evans, N. J., Shirley, Y. L., Mueller, K. E., and Knez, C. (2002). In *Hot Star Workshop III: The Earliest Stages of Massive Star Birth. ASP Conference Proceedings, Vol. 267. Edited by Paul A. Crowther.*, pages 17.

Garay, G., Brooks, K. J., Mardones, D., and Norris, R. P. (2003). *ApJ*, 587:739–747.

Garay, G., Faundez, S., Mardones, D., et al., (2004). *ApJ*, 610:313–319.

Garay, G. and Lizano, S. (1999). *PASP*, 111:1049–1087.

Gibb, A. G., Hoare, M. G., Little, L. T., and Wright, M. C. H. (2003). *MNRAS*, 339:1011–1024.

Henning, T., Schreyer, K., Launhardt, R., and Burkert, A. (2000). *A&A*, 353:211–226.

Hogerheijde, M. R. and Sandell, G. (2000). *ApJ*, 534:880–893.

Hunter, T. R. (1997). *Ph.D. Thesis*.

Jijina, J. and Adams, F. C. (1996). *ApJ*, 462:874.

Keto, E. (2003). *ApJ*, 599:1196–1206.

Konigl, A. (1999). *New Astronomy Reviews*, 43: 67–77.

Konigl, A. and Pudritz, R. E. (2000). *Protostars and Planets IV*, pages 759.

Kurtz, S., Cesaroni, R., Churchwell, E., et al., (2000). *Protostars and Planets IV*, pages 299.

Kurtz, S., Churchwell, E., and Wood, D. O. S. (1994). *ApJS*, 91:659–712.

Kylafis, N. D. and Pavlakis, K. G. (1999). In *NATO ASIC Proc. 540: The Origin of Stars and Planetary Systems*, pages 553.

McKee, C. F. and Tan, J. C. (2002). *Nature*, 416:59–61.

McKee, C. F. and Tan, J. C. (2003). *ApJ*, 585:850–871.

Minier, V., Booth, R. S., and Conway, J. E. (2000). *A&A*, 362:1093–1108.

Molinari, S., Brand, J., Cesaroni, R., and Palla, F. (1996). *A&A*, 308:573–587.

Molinari, S., Brand, J., Cesaroni, R., and Palla, F. (2000). *A&A*, 355:617–628.

Motte, F., Andre, P., and Neri, R. (1998). *A&A*, 336:150–172.

Mueller, K. E., Shirley, Y. L., Evans, N. J., and Jacobson, H. R. (2002). *ApJS*, 143:469–497.

Norberg, P. and Maeder, A. (2000). *A&A*, 359:1025–1034.

Norris, R. P., Byleveld, S. E., Diamond, P. J., et al., (1998). *ApJ*, 508:275–285.

Ossenkopf, V. and Henning, T. (1994). *A&A*, 291:943–959.

Patel, N. A., Greenhill, L. J., Herrnstein, J., et al., (2000). *ApJ*, 538:268–274.

Pestalozzi, M. R., Elitzur, M., Conway, J. E., and Booth, R. S. (2004). *ApJ*, 603:L113–L116.

Richer, J. S., Shepherd, D. S., Cabrit, S., Bachiller, R., and Churchwell, E. (2000). *Protostars and Planets IV*, pages 867.

Ridge, N. A. and Moore, T. J. T. (2001). *A&A*, 378:495–508.

Salpeter, E. E. (1955). *ApJ*, 121:161.

Scalo, J. (1998). In *ASP Conf. Ser. 142: The Stellar Initial Mass Function (38th Herstmonceux Conf.)*, pages 201.

Schilke, P., Groesbeck, T. D., Blake, G. A., and Phillips, T. G. (1997). *ApJS*, 108:301.

Shepherd, D. S. and Churchwell, E. (1996a). *ApJ*, 472:225.

Shepherd, D. S. and Churchwell, E. (1996b). *ApJ*, 457:267.

Shepherd, D. S., Watson, A. M., Sargent, A. I., and Churchwell, E. (1998). *ApJ*, 507:861–873.

Shepherd, D.S., Yu, K. C., Bally, J., and Testi, L. (2000) *ApJ*, 535:833.

Shepherd, D.S., Claussen, M.J., & Kurtz, S.E. (2001) Science, 292:1513–1518.

Shepherd, D. (2003). In *ASP Conf. Ser. 287: Galactic Star Formation Across the Stellar Mass Spectrum*, pages 333–344.

Shepherd, D. S., Testi, L., and Stark, D. P. (2003). *ApJ*, 584:882–894.

Shepherd, D. S., Kurtz, S. E., and Testi, L. (2004). *ApJ*, 601:952–961.

Shirley, Y. L., Evans, N. J., Young, K. E., et al., (2003). *ApJS*, 149:375–403.

Shu, F. H., Najita, J. R., Shang, H., and Li, Z.-Y. (2000). *Protostars and Planets IV*, pages 789.

Sollins, P. K., Hunter, T. R., Battat, J., et al., (2004). *ApJ*, 616:L35–L38.

Sridharan, T. K., Beuther, H., Schilke, P., et al., (2002). *ApJ*, 566:931–944.

Stahler, S. W., Palla, F., and Ho, P. T. P. (2000). *Protostars and Planets IV*, pages 327.

Su, Y., Zhang, Q., and Lim, J. (2004). *ApJ*, 604:258–271.

Torrelles, J. M., Gomez, J. F., Rodriguez, L. F., et al., (1998). *ApJ*, 505:756–765.

Torrelles, J. M., Patel, N. A., Anglada, G., et al., (2003). *ApJ*, 598:L115–L119.

Torrelles, J. M., Patel, N. A., Gomez, J. F., et al., (2001). *Nature*, 411:277–280.

Walmsley, M. (1995). In *Revista Mexicana de Astronomia y Astrofisica Conference Series*, pages 137.

Walsh, A. J., Burton, M. G., Hyland, A. R., and Robinson, G. (1998). *MNRAS*, 301:640–698.

Williams, S. J., Fuller, G. A., and Sridharan, T. K. (2004). *A&A*, 417:115–133.

Wilson, T. L., Gaume, R. A., Gensheimer, P., and Johnston, K. J. (2000). *ApJ*, 538:665–674.

Wood, D. O. S. and Churchwell, E. (1989). *ApJ*, 340:265–272.

Wu, Y., Wei, Y., Zhao, M., et al., (2004). *A&A*, 426:503–515.

Wu, Y., Zhagn, Q., Chen, H., et al., (2005). *ApJ in press*.

Yorke, H. W. and Sonnhalter, C. (2002). *ApJ*, 569:846–862.

Zhang, Q., Hunter, T. R., Brand, J., et al., (2001). *ApJ*, 552:L167–L170.

Paola Caselli, Eric Keto, Charlie Lada, Phil Myers, Ray Jay and Daniel Folha

# OBSERVATIONS OF ACCRETION ONTO HIGH MASS STARS

Eric Keto
*Harvard-Smithsonian Center for Astrophysics*
*60 Garden St., Cambridge, MA, 02138, USA*
keto@cfa.harvard.edu

**Abstract** Observations of the H66$\alpha$ recombination line from the ionized gas in the cluster of newly formed massive stars, G106.–0.4, show that most of the continuum emission derives from the dense gas in an ionized accretion flow that forms an ionized disk or torus around a group of stars in the center of the cluster. The inward motion observed in the accretion flow suggests that despite the equivalent luminosity and ionizing radiation of several O stars, neither radiation pressure nor thermal pressure has reversed the accretion flow. The observations indicate why the radiation pressure of the stars and the thermal pressure of the HII region are not effective in reversing the accretion flow.

## Introduction

Early calculations of stellar evolution indicated that more massive stars have very short pre-main sequence (PMS) lifetimes. For example, Iben (1965) estimated that a 15 $M_\odot$ star has a PMS lifetime of $6 \times 10^4$ yrs versus $5 \times 10^7$ yrs for a 1.0 $M_\odot$ star. The short time scale suggests that unless a massive star forms with a very high accretion rate, $\dot{M} > M_{star}/t_{PMS} \sim 10^{-2}\ M_\odot\ yr^{-1}$, the star will begin burning hydrogen while still in the accretion phase. While Iben's calculations were for stars of a fixed mass, Stahler (1980) and Palla and Stahler (1993) calculated the PMS evolution of stars growing by accretion and confirmed that stars greater than 8 $M_\odot$ have no PMS phase. Beech & Mitalas (1994), Norberg & Maeder (2000), and Behrend & Maeder (2001), extended these results for stars up to 60 $M_\odot$ for various accretion rates.

An example calculation (provided by Alessandro Chieffe based on the models in Chieffi, Staniero, & Salaris 1995) of the evolution of an accreting star starting from an initial mass of 1.0 $M_\odot$ is shown in figure 1. The evolutionary tracks for different accretion rates intersect the main sequence line at about 4 $M_\odot$ and follow that line as the star gains mass. These tracks indicate that a star that is gaining mass by accretion would appear little different in luminosity and temperature from a main sequence star of fixed mass. Thus the stellar structure calculations suggest that if massive stars form by accretion at rates $\dot{M} < 10^{-2}$ M$_\odot$ yr$^{-1}$, then the precursors to massive stars would not resemble massive protostars in a PMS phase. Rather, the precursors to massive stars would appear similar to main sequence A or B stars except that because of the ongoing accretion, they would likely have the characteristics associated with accretion. For example, they would be deeply embedded, show evidence for disks and bipolar outflows, and possibly show maser emission.

However, if massive stars grow by accretion from lower but still high mass main sequence stars, one may ask whether the radiation pressure and thermal pressure generated by the intense main sequence luminosity of the precursor star, and its short main sequence lifetime, might limit the growth of a star by accretion. If the outward forces are too large or the timescale too short, growth by accretion could be limited to a low stellar mass with a spectral type much later than O or B. If so, the most massive stars might not form by accretion.

Radiation pressure from the combined luminosity of the star and the deceleration of the accretion flow at the surface of the star could be enough to reverse the accretion flow (Larson & Starrfield 1969, Yorke & Krugel 1977). Beech & Mitalas (1994) calculated that the most naive comparison between the outward force of the radiation pressure and the inward force of gravitational attraction, $F_{rad}/F_{grav} = [\kappa L/(4\pi r^2 c)]/(GM/r^2)$ indicates that stars of mass $> 8$M$_\odot$ have sufficient luminosity to counter the gravitational attraction driving accretion. Here $\kappa$ is the dust opacity, $L$ the luminosity, $r$ the radius that cancels out of the equation, $c$ the speed of light, and $G$ the gravitational constant. Although this would seem to imply that stars of mass greater than 8 M$_\odot$ could not form by accretion, the comparison is too naive. Kahn (1974) and Wolfire & Cassinelli (1987) showed that if the dust is destroyed out to some radius from the star, then if the accretion flow has sufficient momentum to push the dust grains into the zone of destruction, the accretion flow will be pushed into the dust free zone that is largely transparent to radiation. Nakano (1989) suggested that the flattening of the accretion flow, as would be caused by rotation or magnetic fields, would allow thermal re-radiation from the dust grains to escape out the poles thus reducing the radiation pressure behind the ionization front. The same reduction in pressure will occur if the accretion flow is clumpy and the radiation can escape between the clumps. Jijina & Adams (1996) suggested that the radiation pressure in a rotating accretion flow would first cause the gas to fall onto the accretion disk at larger radii rather than immediately reversing the flow as in the case of spherical accretion. These several more detailed calculations suggest that radiation pressure might be a limiting factor in high mass star formation, but the limiting mass should be larger than 8 M$_\odot$.

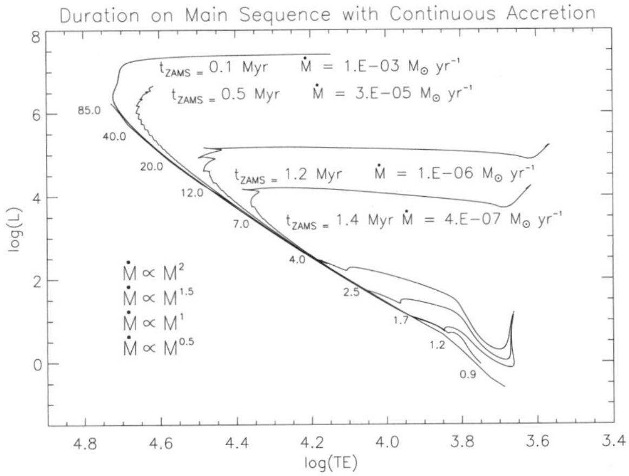


*Figure 1.* HR-diagram with 4 evolutionary tracks for stars gaining mass by accretion at different rates. $t_{\rm ZAMS}$ indicates the time spent evolving up the ZAMS line from a mass of 4 $M_\odot$ star to the mass at which the star deviates off the ZAMS line. The accretion rate at the time of deviation is shown for each track.

The thermal pressure of an HII region could reverse the accretion flow (Nakano, Hasegawa & Norman 1995). Stars of spectral type B and earlier produce significant radiation shortward of the Lyman continuum limit that will ionize the gas around the star. Because the temperature of the ionized gas is on the order of $10^4$ K while the temperature of the molecular gas is of order 100 K, the pressure of the ionized gas will be 100 times greater than in the surrounding molecular gas, and the HII region would expand outward reversing the accretion flow. This might imply that stars of spectral type B would be unable to grow by accretion once an HII region formed around the star. Keto (2002b) suggested that the increased pressure of the ionized gas would be irrelevant to the accretion flow as long as the zone of ionization were within the sonic radius where the escape speed from the star exceeded the sound speed of the ionized gas. This would occur if the gas in the accretion flow were dense enough that the radius of ionization equilibrium, where recombination balances the ionization, were smaller than the sonic radius.

The short main sequence lifetime of high mass stars compared with the accretion timescale might be the limiting factor in high mass star formation (Nakano 1989, Bernasconi & Maeder 1996, Behrend & Maeder 2001). If the accretion rate is not high enough to supply sufficient fresh hydrogen to dilute the helium ash of hydrogen burning, then the star will evolve off the main sequence where stellar pulsations and the eventual ejection of the star's envelope will disrupt and end the accretion onto the star. Figure 1 shows that the accretion rate has to be quite high $\dot{M} > 10^{-3}$ $M_\odot$ $yr^{-1}$ to keep a growing star on the main sequence long enough to reach the mass of an O star. Such a high rate might be the result of accretion that increases as some function of the mass of the star as indicated in the figure.

Observations of accretion flows around high mass stars may provide some guidance toward understanding how high mass stars may form by accretion despite the three limiting effects of radiation pressure, thermal pressure, and the short main sequence lifetime of high mass stars. Because most of the massive stars, with a few exceptions such as the Orion star forming region, are at kpc distances, it is difficult to obtain the sensitivity and angular resolution to measure the properties of accretion flows around individual high mass stars. Instead we can measure accretion flows around small groups of a few stars. This mode of formation may be common in high mass star formation. For example, observations by Prebisch et al. (2001) suggest that massive stars form in binaries and small groups at a greater frequency than do low mass stars.

We chose as a target for our observations the accretion flow onto a small group of stars at the center of the G10.6–0.4 cluster (Keto 2002a, Sollins et al. 2004). This source is appropriate for our study because the combined mass of the group, higher than for a single star, generates accretion velocities that are high enough and on size scales that are large enough that we can measure the properties of the flow.

## Observations

The new observations were made with the Very Large Array (VLA) radio interferometer on February 23, 2002. The array was in the "A" configuration with baselines up to 36 km leading to an angular resolution of $0.2 \times 0.1$ arcsec. The correlator was configured for a bandpass of 12.5 MHz with 64 channels of 195.31 kHz or 2.62 kms$^{-1}$ at the observing frequency of the H66$\alpha$ recombination line of hydrogen at 22.36416932 GHz. The data were combined with H66$\alpha$ observations made in the D configuration and described separately in Keto (2002a) and images and spectra made with the combined data set. The sum of the channels off the spectral line yields a map of the continuum emission that when subtracted from the total emission yields a map of the H66$\alpha$ spectra.

## The Accretion Flow

Figures 4 and 5 show the continuum emission around the group of stars. The plane of the accretion disk is indicated in figure 4 by the solid line at 45° across the figure. The orientation of the bipolar outflow perpendicular to the disk is shown by the dotted line. Figure 5 is a detail of photo-ionized arcs to the left of the bright HII region. The location is marked as "ARCS" in figure 4.

The bright continuum emission seen in the new high angular resolution observations of G10.6–0.4 is identified as a dense accretion torus rather than a classical ionization bounded HII region by the orientation of the emission with respect to the accretion flow and by the presence of the nearby photo-ionized arcs. Previous observations of molecular emission and absorption in G10.6–0.4 have mapped the structure of the molecular accretion flow onto the stars and identified the axis of rotation as projected onto a northeast-southwest direction (Keto 1990, Sollins et al. 2004). If the continuum emission were from a classical HII region with a boundary defined by equilibrium between ionization and recombination, the HII region would be extended down the density gradient in the direction of the lower density gas, perpendicular to the rotational flattening of the accretion flow. In this case the continuum emission would have an elliptical shape elongated in the direction of the rotation axis. However, the observations, figure 4, show that the continuum emission is seen elongated along the equatorial plane following the densest gas of the accretion flow.

The absence of an ionization boundary around the bright continuum emission is also indicated by the photo-ionized arcs most prominently on the southeast (left) side of the central ellipse, figures 4 and 5. The orientation of the photo-ionized arcs with their ionization fronts facing the stars indicates that these outlying clumps of high density molecular gas receive ionizing radiation from the stars. This would not be possible if there were an ionization front at the boundary of the bright continuum emission. The gas between the bright continuum ellipse and the photo-ionized arcs must be ionized to allow the stellar radiation to pass and reach the arcs, and the low continuum brightness between the arcs and the disk confirms that this gas has a lower density. Thus the apparent boundary of the bright continuum ellipse is set by the outwardly decreasing density of the ionized gas rather than by an ionization boundary.

While the bright continuum emission indicates the morphology of the densest ionized gas around the group of stars, the H66$\alpha$ line provides the information on the velocities within the ionized gas to identify this dense gas as an accretion disk or torus. In our data we are also able to identify two other components of the accretion flow, infall in a quasi-spherical envelope and a high velocity outflow aligned with the rotation axis.

Figure 4 shows the trend of velocity with position in the direction along the equatorial plane of the disk, at the position marked by the solid line in figure 2. The backward "C" of emission seen in figure 2 is characteristic of radial or spherical infall. The "C" structure arises because on lines of sight progressively closer toward the center of the infall, the velocity of the emission is more redshifted since the velocity of the radial infall is projected more directly along the line of sight. The "C" shape is characteristic of emission from the front hemisphere of a quasi-spherical inflow. Emission from the back hemisphere would produce a corresponding "C" in the opposite orientation. Optical depth effects are likely responsible for the weakness of the emission from the back of the flow (Keto 2002a). The velocity of the quasi-spherical infall in the ionized envelope is indicated at $> 10$ kms$^{-1}$ and the rotation at $> 10$ kms$^{-1}$.

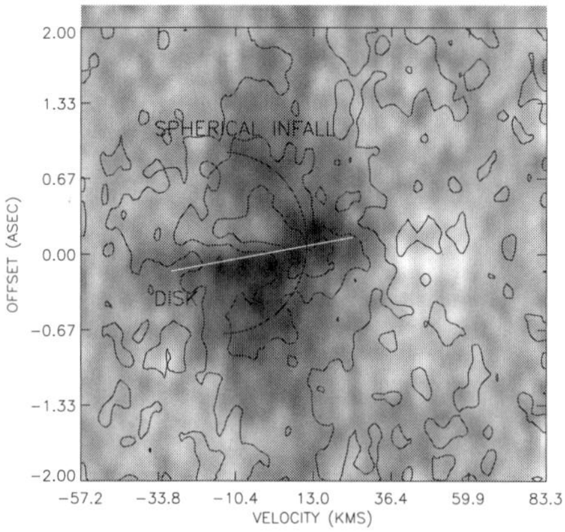


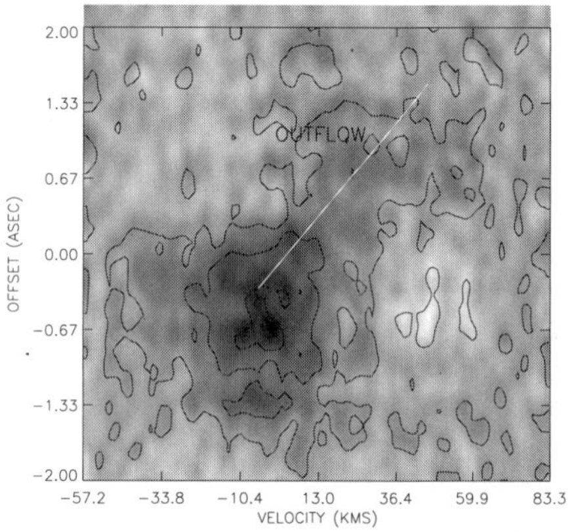


*Figure 2.* Position versus velocity diagram of the H66$\alpha$ recombination line emission from the star cluster G10.6–0.4. The position axis lies on a southeast-northwest line across the long axis of the ionized accretion disk and marked as a solid line in figure 4. The infall in the spherical envelope is seen in the backward "C" structure of the largest scale emission, indicated by the curved black line. The rotating disk is seen in the brighter, smaller scale emission at the center of the "C". The rotation of the disk is seen in the velocity gradient of the brightest emission across the disk. The linear gradient of the rotation is marked by a white line. The spherical infall is extended on a larger scale along the disk than perpendicular to the disk (figure 3) because the scale and the "C" in the perpendicular direction are flattened due to the rotation. The color scale ranges from 0.0 to 0.01 Jy/beam. The contours are in units of 20% of the peak emission of 0.01 Jy/beam.

*Figure 3.* Position versus velocity diagram of the H66$\alpha$ recombination line emission from the star cluster G10.6–0.4. The position axis, marked as a dotted line 4, lies on a southwest-northeast line across the rotation axis and perpendicular to the ionized accretion disk in figure. A high velocity outflow is seen on the northeast side of the disk extending up to 60 kms$^{-1}$ before falling below the sensitivity limit. The location of the outflow is indicated by a white line. The counter outflow on the southwest side of the disk is not seen because the emission there is below the sensitivity limit. The color scale ranges from 0.0 to 0.01 Jy/beam. The contours are in units of 20% of the peak emission of 0.01 Jy/beam.

More rapid rotation of the accretion disk at a smaller scale toward the center of the group of stars is indicated by the breadth of the emission lines in the center of figure 2 and by the linear gradient in line center velocities across the figure, along the plane of the disk. The breadth of the lines derives from a combination of pressure broadening due to the high density of the gas in center of the disk (Keto et al. 1995), and the high rotational velocities that are spatially unresolved. The more rapid rotation toward the center of the disk and the more modest rotation in the envelope together show that the accretion flow is spinning up as the gas spirals in toward the stars.

Figure 3 is a position-velocity diagram of the H66$\alpha$ emission along the axis of rotation through the middle of the ionized disk at the position marked by the dotted line in figure 4. A high velocity outflow of ionized gas is seen in figure 3 in the fainter emission extending northwest from the center of the disk, from a $V_{LSR}$ of near zero kms$^{-1}$ to over 60 kms$^{-1}$ where the recombination line emission falls below the sensitivity limit. Referring back to the map of continuum emission, figure 4, the northeast outflow appears bounded by two lines of continuum emission in a "V" shape. The gas density inside the outflow is quite low as indicated by the weakness of the continuum emission at the location of the outflow. There is a corresponding "V", of wider opening angle, extending to the southwest indicating a counter outflow. The opening angles of both outflows are wider than those of high velocity jets associated with individual high mass stars, and the outflow may be driven by the collective effects, stellar winds or individual jets, of the several massive stars in the central group.

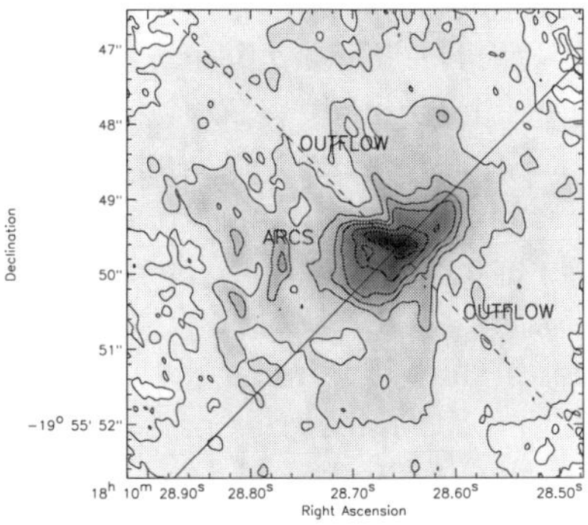


*Figure 4.* Continuum emission at 1.3 cm. Solid line and dotted lines: the disk and rotation axes. Contour levels are 10% of the peak flux of 0.04 Jy/beam.

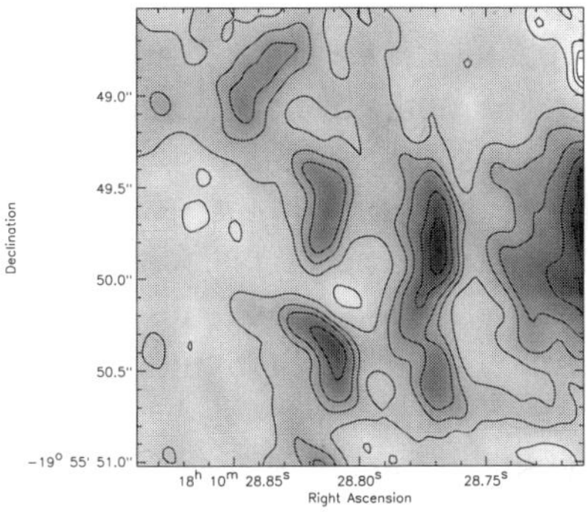


*Figure 5.* Detail of the arcs seen in 1.3 cm emission. Location marked as 'ARCS' on 4. Contour levels are 10% of the peak flux of 0.01 Jy/beam.

## A simple model of the accretion flow

The accretion flow onto G10.6–0.4 is quasi-spherical at larger distances ($>$ 0.1 pc) from the central group of stars (Keto 1990) and, as seen in the current high angular resolution observations, forms an accretion disk on the scale of a few thousand AU. A particularly simple model for an accretion flow with these properties is that of the gas orbiting on ballistic trajectories around a point mass (Ulrich 1976). In this model for steady state accretion, the self-gravity and pressure of the gas are assumed negligible, the point mass at the center of the flow is constant, and the cloud has a specific angular momentum, $\Gamma_\infty$, which is conserved on the flow trajectories. Where the trajectories would cross the mid-plane of the flow, the gas is assumed to collide with a similar flow from the other side of the plane with the gas settling into a disk on the mid-plane. The model produces a reasonable approximation within the gravitational radius of the flow, $r_G = GM/c^2$, where the velocities are supersonic and the gas is nearly in free-fall and away from the mid-plane where the densities in the model are formally infinite. The radius, $r_D$, where the centrifugal force equals the gravitational force, $\Gamma^2/r_D^3 = GM/r_D^2$, divides the flow into an outer region where the inflow is quasi-spherical and an inner region where the flow is rotationally dominated.

Rough estimates for the central group of stars in the G10.6–0.4 cluster are a combined mass of greater than 300 $M_\odot$ determined by the observed infall and rotational velocities, gravitational radius, $r_G = GM/c^2$ greater than 5000 AU, radius of disk formation of 3500 AU, and ionizing luminosity of $2 \times 10^{50}$ $s^{-1}$ photons beyond the Lyman limit (Sollins 2004). The infall and rotational velocities in the overlying molecular flow are similar at about 4.5 $kms^{-1}$ each at a radius of about 5000 AU (Keto 2002a). The infall and rotational velocity components maintain their similarity as the accretion flow spins up approaching the stars, reaching velocities $> 10$ $kms^{-1}$ at a radius of 5000 AU.

The formation of an ionized accretion disk within the ionized portion of an accretion flow is a different model for accretion onto high mass stars than one in which the radius of disk formation is within the molecular portion of the flow. In this latter case the inner part of the molecular accretion disk within the ionized zone would be a photoevaporating disk (Hollenbach et al 1994; Johnstone et al. 1998). The different cases can be distinguished observationally because the ionized gas around a photoevaporating disk derives from the disk and is moving outward away from the disk. The observations of G10.6–0.4 indicate that in this source the disk derives from the surrounding ionized gas and is moving inward toward the disk.

The outflow seen along the rotation axis in G10.6–0.4 must be due to forces not included in the infall model. Owing to the initial conditions of this model, the radial infall velocities are the highest along the polar axis; however, the density gradient is also the steepest. In the context of this model, the breakout of the outflow along the polar axis would be related to the density gradient or possibly to bipolar outflows from individual stars if their individual rotation axes are aligned with the large scale flow onto the stars.

## Implications for star formation by accretion

The new observations of G10.6–0.4 show that the detailed structure of the accretion flow exemplifies several of the factors thought to be important in mitigating the three limiting factors, radiation pressure, thermal pressure, and the short main sequence lifetime of high mass stars.

The observations of G10.6–0.4 show infall despite high luminosity $9.2\times10^5$ $L_\odot$ (Casoli et al. 1986) and radiation pressure $L/c \sim 1 \times 10^{29}$. An estimate of the force deriving from the accretion flow, $\dot{M}v \sim 3 \times 10^{28}$ dynes, based on an accretion rate and velocity of $10^{-3}$ $M_\odot$ yr$^{-1}$ and 4.5 km s$^{-1}$ at 5000 AU (Keto 2002b), suggests that the inward momentum of the accretion flow is of the magnitude required to compete with the radiation pressure. In addition the observations show that the accretion flow is flattened by rotation and that the molecular gas in the flow is clumpy. Both the flattening and clumpiness reduce the effective radiation pressure as suggested by the theoretical models referenced in the introduction. In addition, the radiation exerts negligible pressure on the inner ionized portion of the accretion flow that is transparent to the stellar luminosity.

The observations of G10.6–0.4 show infall directly through the HII region despite the thermal pressure of the hot ionized gas. The simple model for the accretion flow indicates that the zone of ionization of the denser gas is within the gravitational radius of 5000 AU of the combined mass of the stars. Thus the gravitational attraction is sufficient to counter the thermal pressure of the hot ionized gas, and the ionized gas as well as the molecular gas (indicated in these observations by the photoionized arcs) will be moving inward as parts of the accretion flow.

The luminosity of G10.6–0.4 indicates the presence of several O stars (Sollins et al. 2004) that must either have formed in a timescale shorter than their main sequence lifetime or with an accretion high enough to delay their main sequence evolution. The observations indicate a mass accretion rate of about $10^{-3}$ $M_\odot$ yr$^{-1}$ (Keto2002a) on the order of that required to maintain O stars on the main sequence as indicated by the timescales in figure 1.

Thus the new observations of G10.6–0.4 indicate that high mass stars form by accretion and indicate how the accretion flow is able to continue even after the stars have become massive B and O stars. Only the most massive accretion flows have sufficient momentum to counter the radiation pressure of massive stars, the high densities to keep the radius of ionization equilibrium of the HII regions within the sonic radius of the stars, and the mass accretion rate to keep the high mass stars from evolving off the main sequence.

## References

Beech, M., & Mitalas, R.(1994). *ApJS* 95:517

Behrend, A. & Maeder, A. (2001). *A&A* 373:190

Bernasconi, P. & Maeder, A. (1996). *A&A* 307:839

Chieffi, A., Staniero, O., & Salaris, M. (1995). *ApJ* 445:L39

Casoli, F., Combes, F., Dupraz, C., Gerin, M., Boulanger, F. (1986). *A&A* 169:281

Ho, P., & Haschick, A. (1986). *ApJ* 304:501

Hollenbach, D., Johnstone, D., Lizano, S. & Shu, F. (1994). *ApJ* 428:654

Johnstone, D., Hollenbach, D. & Bally, J. (1998). *ApJ* 499:758

Kreugel, E. & Siebenmorgen, R. (1994). *A&A* 288:929

Iben, I. (1965). *ApJ* 141:993

Jijina, J. & Adams, F. (1996). *ApJ* 462:874

Kahn, F. (1974). *A&A* 37:149

Keto, E. (1990). *ApJ* 355:190

Keto, E., Welch, W., Reid, M., Ho, P. (1995). *ApJ* 444:765

Keto, E. (2002a). *ApJ* 568:754

Keto, E. (2002b). *ApJ* 580:980

Keto, E. (2003). *ApJ* 599:1196

Larson, R. & Starrfield, S. (1969). *A&A* 13:190

Nakano, T. (1989). *ApJ* 345:464

Nakano, T., Hasagawa, T. & Norman, C. (1995). *ApJ* 450:183

Norberg, P. & Maeder, A. (2000). *A&A* 359:1025

Palla, F., & Stahler, S. (1993). *ApJ* 418:414

Prebisch, T., Hofmann, K., Schertl, D., Weigelt, G., Balega, Y. Balega, I. & Zinnecker, H. (2001). In *IAU Symp 200. The Formation of Binary Stars, ed: Zinnecker, H. & Mathieu, R., (SanFrancisco: ASP)* 106

Sollins, P., Zhang, Q., Keto, E. & Ho, P. (2004). *ApJ* in press, astro-ph/0410604

Stahler, S., Shu, F. & Taam, R. (1980a). *ApJ* 241:637

Ulrich, R., (1976). *ApJ* 210:377

Wolfire, M. & Cassinelli. (1987). *ApJ* 319:850

Yorke, H., & Krugel, E. (1977). *A&A* 54:183

# DISKS AROUND MASSIVE (PROTO)STARS

Riccardo Cesaroni
*INAF, Osservatorio di Arcetri, Largo Fermi 5, I-50125 Firenze, Italy*
cesa@arcetri.astro.it

**Abstract**

Theory predicts that stars in excess of $\sim 8\ M_\odot$ must form while still accreting material from the parental cloud: at this stage radiation pressure should reverse the infall thus preventing further growth of the stellar mass. After illustrating the two models proposed to solve this problem ("accretion" and "coalescence"), we review the observational evidence pro/contra such models, focusing on rotating disk-like structures detected in association with luminous young (proto)stars. Two types of such objects have been found: less massive, Keplerian disks rotating about a single star or binary system; and more massive "toroids" probably enshrouding a cluster of stars. The existence of such disks/toroids seems to favour the accretion model for high-mass stars.

## The dividing line between high- and low-mass stars

When dealing with the formation of stars, the term "massive" is commonly used to identify stars with spectral types ranging from early B to O. In practice, the dividing line is set at a value of $\sim 8\ M_\odot$ (Palla & Stahler 1993). Such a definition is related to the fact that the mode of formation of a star seems to depend dramatically on the stellar mass. Let us illustrate this result in some better detail.

In a widely accepted scenario, the formation of low-mass stars proceeds through inside-out collapse of an isothermal core (Shu et al. 1987). In a simplified description of the process, one may imagine that accretion proceeds onto a dense kernel, the "protostar", which radiates the gravitational energy released in the collapse. To this process two timescale are relevant: the accretion timescale of the envelope, $t_{\rm acc} = M_*/\dot{M}_{\rm acc}$, and the contraction timescale of the kernel, the so called Kelvin-Helmotz timescale, $t_{\rm KH} = GM_*/R_*L_*$. Here we have indicated with $M_*$, $R_*$, and $L_*$ respectively the mass, radius, and luminosity of the protostar, and with $\dot{M}_{\rm acc}$ the mass accretion rate. As long as $M_* < 8\ M_\odot$, $t_{\rm KH}$ exceeds $t_{\rm acc}$, whereas for for larger masses the reverse occurs, i.e. $t_{\rm KH} < t_{\rm acc}$ (Palla & Stahler 1993). This has the consequence that stars below 8 $M_\odot$ contract very slowly as compared to accretion, so that, once their reservoire of material is exhausted, they may slowly contract and heat up: they eventually ignite hydrogen burning, and reach the zero-age main sequence (ZAMS). On the other hand, if the reservoire (i.e the parental clump) is large enough, the star will keep growing until the critical mass of 8 $M_\odot$ is reached. Beyond this point it will rapidly collapse (because $t_{\rm KH} < t_{\rm acc}$) thus reaching the ZAMS still deeply embedded.

At this stage, the radiation pressure and copious UV flux emitted by an O–B star are bound to halt the collapse and disrupt the surrounding molecular cloud, thus preventing further growth of the stellar mass. The paradoxical conclusion is that massive stars (i.e. stars above $\sim$8 $M_\odot$) should not form. Such a paradox clearly demonstrates that formation theories successfully applied to low-mass stars cannot be trivially extended to their high-mass counterparts.

## Possible scenarios for high-mass star formation

Eventually, what hinders further growth of a massive protostar is the radiation pressure exerted on dust grains in the infalling material. The problem has been clearly illustrated by Yorke (2004): to make accretion possible, the dust absorption coefficient, $\kappa_{\rm dust}$, must be less than a critical value proportional to the ratio $M_*/L_*$. This result may be obtained by reducing $\kappa_{\rm dust}$ or by overcoming the radiation pressure with the ram pressure of the infalling gas. Non-spherical accretion is very effective for this purpose, as it is much more difficult for the radiation to blow away a massive disk. Basically, two models have been proposed to solve the problem of massive star formation: *(non-spherical) accretion* with sufficiently large accretion rates (Behrend and Maeder, 2001; Yorke and Sonnhalter, 2002; Tan and McKee, 2002); and *coalescence* of stars with masses below the critical value of $\sim$8 $M_\odot$ (Bonnel & Bate 2002). As pointed out by Yorke (2004), the latter may be seen as an extreme case of accretion with low values of $\kappa_{\rm dust}$, as radiation pressure is totally ineffective against merging stars.

It is helpful to compare the implications of the two models and see which of these is supported by the observations. Clearly, the most distinctive feature is the fact that isolated massive star formation is impossible in the coalescence model, for which the presence of a large number of stars ($\sim$ $10^8$ pc$^{-3}$) is mandatory, whereas it can be explained in the accretion scenario. However, the latter predicts that high-mass stars form at the centre of massive molecular clumps which give origin also to a large number of lower mass stars, thus making isolated star formation very unlikely in any case. A much more effective way to discriminate between the two models is given by the kinematics of the gas associated with the forming stars. In fact, both scenarios require

*Table 1.* List of tracers used to search for disks in high-mass YSOs and corresponding references

| Tracer | References |
|---|---|
| $CH_3OH$ masers | Norris et al. (1998); Phillips et al. (1998); Minier et al. (1998;2000); Pestalozzi et al. (1994) |
| OH masers | Cohen et al. (2004); Edris et al. (2005) |
| SiO masers | Greenhill (2004) |
| $H_2O$ masers | Fiebig et al. (1996); Torrelles et al. (1996) ; Shepherd and Kurtz (1999) |
| NIR, mm, cm continuum | Gibb et al. (2004) |
| $NH_3$, $C^{18}O$, CS, $C^{34}S$ $CH_3CN$ | Keto et al. (1988); Cesaroni et al. (1997; 1998; 1999); Shepherd and Kurtz (1999); Shepherd et al. (2000); Sandell et al. (2003); Beltrán et al. (2004); Zhang et al. (1998a; 1998b; 2002); |

the presence of infalling gas which, by conservation of angular momentum, is bound to form flattened, rotating structures, i.e. circumstellar disks. The accretion model is consistent with the existence of well defined disks. In the coalescence model, instead, the disks associated with the low mass stars could be destroyed during the merging process. It is hence clear that the detection of disks in high-mass YSOs would lend support to the accretion model.

## Evidence for disks in massive (proto)stars

In view of the previous considerations, it is of great interest to review the observational evidence for rotating disks in massive YSOs. While disks have been directly observed in a large number of low-mass YSOs (Simon et al. 2000), evidence for rotating disks in massive star forming regions is provided by a limited number of studies. These are summarised in Table 1, where the tracers used and the corresponding references are reported. As one may see, there are basically three different types of tracers: maser lines, continuum emission, and thermal lines. Let us discuss these in some better detail.

## Disks traced by maser lines

Maser lines are the most suitable to obtain a high angular resolution picture of the circumstellar environment, as they arise from regions as small as a few AU with brightness temperatures which may reach $10^{12}$ K. These characteristics make them ideal targets for VLBI studies and excellent tracers of the velocity field. However, other problems hinder the use of maser emission: the large time variability of some masers (especially $H_2O$) and the fact that they often arise in shocked regions, make their observation and interpretation quite difficult and often ambiguous. Even in the best studied cases such as that of

Orion KL (Greenhill et al. 2004), where a 3-D view of the velocity field has been obtained with proper motion measurements of the SiO maser spots, it remains unclear whether one is observing rotation, outflow, or infall, or possibly a combination of these. Similarly, $H_2O$ masers have also given ambiguous results. For instance, in Cep A HW2 (Torrelles et al. 1996) the $H_2O$ maser spots are distributed in an elongated structure perpendicular to the axis of a thermal jet arising from a luminous YSO. However, a velocity gradient in the spots is seen both in the direction of the axis and perpendicular to it. It is hence unclear whether one is observing rotation about the jet axis, expansion along that axis, or infall towards the YSO.

$CH_3OH$ masers are also ambiguous. In many cases, the maser spots turn out to be distributed along lines or arcs with velocity monotonically increasing along these structures. Several authors (see Table 1) have interpreted these as Keplerian disks. Under this assumption one may obtain a lower limit to the mass of the central (proto)star from the measured velocity gradient. Although for a few objects this method gives reasonable values for the stellar mass (e.g. 6 $M_\odot$ in W48; see Minier et al. 2000), in most cases the masses obtained in this way are less than 1 $M_\odot$. Moreover, De Buizer (2003) has searched for jets in the 2.2 $\mu$m line towards a sample of $CH_3OH$ maser sources claimed to trace rotating disks. In all cases where a jet could be identified, the direction of this is parallel to the maser structures, and not perpendicular to them as expected if the masers were tracing disks.

A general conclusion concerning masers as tools to search for disks in massive YSOs, is that they can be of limited use without additional information on the source, such as e.g. the direction of an associated jet/outflow. At present, while convincing evidence has been reported on the association between masers and outflows (see e.g. Moscadelli et al. 2000) the association between masers and disks in YSOs remains questionable.

## Disks traced by continuum emission

In low-mass star forming regions, thermal line and continuum emission have been successfully used to trace circumstellar or circumbinary disks, the first and perhaps most striking case being that of GG Tau Guilloteau et al. ( 2002). Since continuum emission at (sub)mm wavelengths is proportional to the mass and temperature of the dust (and hence of the gas, given the coupling between the two at large densities), this should be even more effective for massive/luminous YSOs, which one might naïvely expect to be associated with more massive and hotter disks. However, in this case typical distances are an order of magnitude larger than for low-mass YSOs, which makes it very difficult to resolve the disk and disentangle its emission from that of a possible stellar wind. An example of searches based on mm continuum emission

*Table 2.* List of disks and toroids in high-mass star forming regions

| Name | $L(L_\odot)$ | $M(M_\odot)$ | $R$(AU) | $M_*(M_\odot)$ | Ref.[a] |
|---|---|---|---|---|---|
| G192.16–3.82 | $3\,10^3$ | 15 | 1000 | 6–10 | 1,2 |
| AFGL 5142 | $4\,10^3$ | 4 | 1800 | 12 | 3 |
| G92.67+3.07 | $5\,10^3$ | 12 | 14400 | 4–7.5 | 4 |
| IRAS 20126+4104 | $10^4$ | 4 | 1600 | 7 | 5,6,7,8 |
| M17 | ? | >110 | 20000 | 15–20 | 9 |
| NGC7538 S | $10^4$ | 100–400 | 30000 | 40 | 10 |
| G29.96–0.02 | $9\,10^4$ | 300 | 14000 | <25[b] | 11 |
| G31.41+0.31 | $3\,10^5$ | 490 | 16000 | <40[b] | 12,13 |
| G24.78+0.08[c] | $7\,10^5$ | 80–250 | 4000–8000 | 20–40[b] | 13 |

[a] 1: Shepherd and Kurtz (1999); 2: Shepherd et al. (2001); 3: Zhang et al. (2002); 4: Bernard et al. (1999); 5: Cesaroni et al. (1997); 6: Zhang et al. (1998b); 7: Cesaroni et al. (1999); 8: Cesaroni et al. (2005); 9: Chini et al. (2004); 10: Sandell et al. (2003); 11: Olmi et al. (2003); 12: Cesaroni et al. (1994); 13: Beltrán et al. (2004).
[b] upper limit assuming that luminosity originates from one ZAMS star.
[c] 3 toroids have been found in the G24.78 star forming region.

is that of Gibb et al. (2004), who also combined the information from their BIMA observations of a sample of massive YSOs, with previous NIR images (Hoare et al. 1996). Although suggestive of the presence of circumstellar disks, their findings confirm that better angular resolution and sensitivity at mm wavelengths are needed to draw a firm conclusion.

## Disks traced by thermal line emission

The obvious advantage of observations of thermal lines with respect to continuum is that these supply us with information on the velocity of the gas. This makes it possible to reveal a rotating disk through its velocity gradient even if the angular resolution may be insufficient to properly resolve its structure. With millimeter interferometers one may cover a large number of lines from different molecular species, which is of great importance to investigate the kinematics of the molecular cloud on different spatial scales. This is especially important when searching for disks. In fact, the detection of a velocity gradient in a core does not by itself guarantee the existence of a rotating disk; but if such a gradient turns out to be perpendicular to the axis of an outflow, the disk interpretation becomes very likely. Indeed accretion disks and outflows are believed to be tightly associated: with this in mind, detailed analyses of massive YSOs with outflows have been performed through interferometric observations of core *and* outflow molecular tracers. Once a velocity gradient in the core is observed, the information on the outflow direction is sufficient by itself to establish whether the gradient in the core is due to rotation or instead is just tracing the root of the outflow itself: in the latter case, the gradient in the core should be parallel to the outflow axis, whereas rotation requires perpendicularity between the two.

This technique has been applied successfully in several luminous YSOs, leading to the discovery of a handful of rotating disks. These are reported with their properties in Table 2, sorted according to the bolometric luminosity. It must be stressed that in most cases it is unclear whether the luminosity (which comes from IRAS measurements) is indeed originating from a single YSO or a whole cluster: nevertheless, higher luminosities indicate significant differences in the properties of the region observed and hence of the disk associated with it, no matter whether one is observing a circumstellar or a "circumcluster" disk. As a matter of fact, looking at Table 2 one can see that B (proto)stars are associated with less massive and smaller disks whereas the opposite occurs with more luminous YSOs. With this in mind, in the following disks below a few 10 $M_\odot$ will be referred to as "disks", whereas the term "toroids" will be used for the others. The main reason for using a different terminology is related to the ratio between the mass of the disk and that of the star: while this ratio is around unity for B (proto)stars, in more luminous objects it becomes $>10$, which proves the disk to be dynamically unstable. We thus believe that a distinct terminology is in order for such "fat" disks. In Figs. 1 and 2 prototypical examples of a disk and a toroid are shown. The former illustrates the structure of IRAS 20126+4104 from $\sim$0.2 pc to $\sim$200 AU. On all scales a bipolar jet/outflow oriented SE–NW is clearly seen in a variety of tracers such as SiO, $H_2$, 3.6 cm continuum, and $H_2O$ masers. At the geometrical centre of this flow a compact 7 mm continuum source has been detected with the VLA (Hofner, pers. comm.): most likely this marks the position of a luminous YSO deeply embedded in the molecular core seen in the mm continuum with the Plateau de Bure interferometer by Cesaroni et al. (1997; 1999). These authors have also mapped the core in $CH_3CN$, a high density tracer, revealing a velocity gradient perpendicular to the jet/outflow axis, which they interpret as a Keplerian disk rotating about the embedded YSOs. Recent observations in the $C^{34}S$(2–1) and (5–4) lines (Cesaroni et al. 2005) confirm this hypothesis, leading to an estimate of $\sim$7 $M_\odot$ for the mass of the star, and in addition provide us with evidence of infall in the core. We conclude that in IRAS 20126+4104 one is dealing with a circumstellar accretion disk in Keplerian rotation about an early-B (proto)star.

The situation in the case of G24.78+0.08 is significantly different from that of IRAS 20126+4104. Figure 2 illustrates a comparison between the bipolar outflows detected by Furuya et al. (2002) with the velocity gradients found by Beltrán et al. (2004) in the cores hosting the YSOs powering the flows. This picture is reminiscent of Fig. 1, but in G24.78+0.08 the luminosity and mass of the toroids are almost two orders of magnitude greater than in IRAS 20126+4104. Clearly, Keplerian rotation is impossible as the disk mass is much larger than

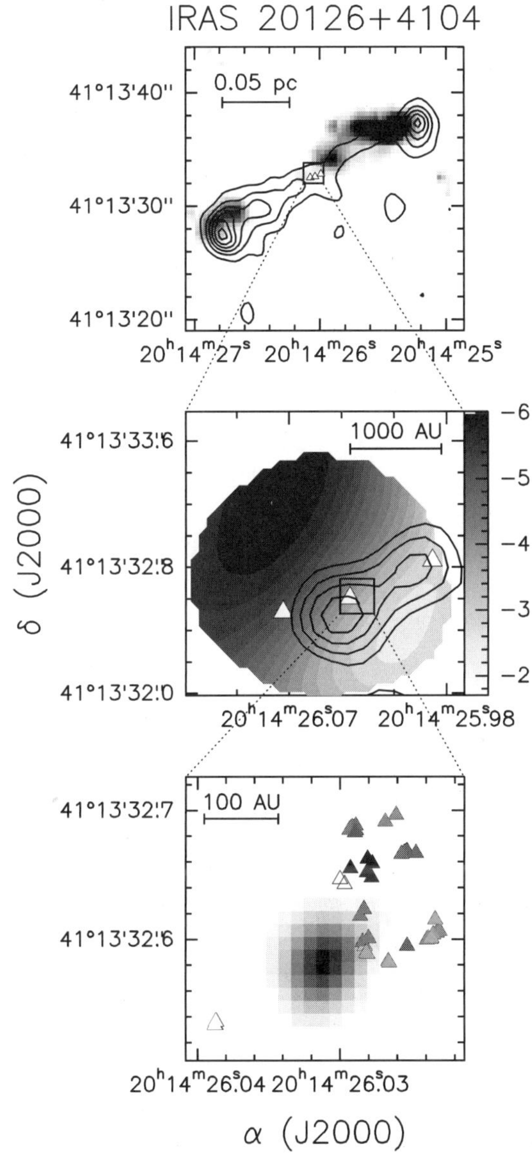


*Figure 1.* Disk/outflow system in the high-mass (proto)star IRAS 20126+4104. Top: Overlay of the $H_2$ lime emission at 2.2 $\mu$m (grey scale) and the jet map obtained from the SiO(2–1) line (Cesaroni et al. 1997). The triangles mark the positions of the $H_2O$ maser spots detected by Tofani et al. (1995). Middle: 3.6 cm continuum map (contours; Hofner et al. 1999b) overlayed on a map of the velocity measured in the $C^{34}S(5–4)$ line by Cesaroni et al. (2005). Bottom: Distribution of the $H_2O$ maser spots (Moscadelli et al. 2000; Moscadelli et al. submitted) compared to a VLA map (image) of the 7 mm continuum emission (Hofner pers. comm.). The grey scale of the spots ranges from white, for the most red-shifted spots, to black, for the most blue-shifted.

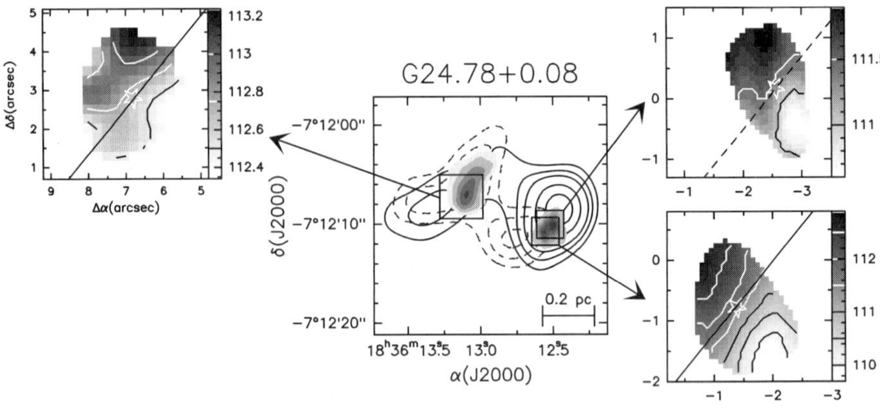


*Figure 2.* Disk/outflow systems in the high-mass star forming region G24.78+0.08 (Beltrán et al. 2004). The straight lines represent the outflow axes. Left: Map of the CS(3–2) line velocity observed by Cesaroni et al. (2003) towards G24 C. The grey scale is expressed in km s$^{-1}$. Middle: Comparison between the bipolar outflows observed by Furuya et al. (2002), the $CH_3CN$(12–11) line emission map towards G24 A1 and G24 A2, and the CS(3–2) line map Furuya et al. (2002) towards G24 C. Right: Map of the $CH_3CN$(12–11) line velocity towards G24 A1 and G24 A2.

any reasonable stellar mass. Also, Beltrán et al. (2004) find that the masses of the toroids are significantly greater than the dynamical masses required for equilibrium and conclude that one is observing unstable, infalling/rotating structures.

A general conclusion that may be drawn from all these findings is that, while in B (proto)stars stable (possibly Keplerian) disks are seen, in more luminous regions only very massive, unstable toroids are found, very likely rotating about a cluster of O–B (proto)stars. One may refer to these as "circumcluster" toroids. It must be noted that such toroids although unstable, cannot be short-lived as they have been detected in several sources: this implies that they must be continuously supplied with fresh material accreting from the surrounding molecular envelope.

## Conclusions

Observations of high-mass star forming regions have revealed the existence of rotating circumstellar, Keplerian disks or circumcluster, massive toroids. The evidence collected so far lends support to the hypothesis expressed by various authors (e.g. Garay and Lizano 1999) that also high-mass stars may form through non-spherical accretion as much as their low-mass counterparts, although with larger mass accretion rates.

## References

Behrend, R. and Maeder, A.: 2001, *A&A* **373**, 190.

Beltrán, M.T., Cesaroni, R., Neri, R., Codella, C., Furuya, R.S., Testi, L., and Olmi, L.: 2004, *ApJ* **601**, L187.

Bernard, J.P., Dobashi, K., and Momose, M.: 1999, *A&A* **350**, 197.

Bonnell, I.A. and Bate, M.R.: 2002, *MNRAS* **336**, 659.

Cesaroni, R., Olmi, L., Walmsley, C.M., Churchwell, E., and Hofner, P.: 1994, *A&A* **435**, L137.

Cesaroni, R., Felli, M., Testi, L., Walmsley, C.M., and Olmi, L.: 1997, *A&A* **325**, 725.

Cesaroni, R., Hofner, P., Walmsley, C.M., and Churchwell, E.: 1998, *A&A* **331**, 709.

Cesaroni, R., Felli, M., Jenness, T., Neri, R., Olmi, L., Robberto, M., Testi, L., and Walmsley, C.M.: 1999, *A&A* **345**, 949.

Cesaroni, R., Codella, C., Furuya, R.S., and Testi L.: 2003, *A&A* **401**, 227.

Cesaroni, R., Neri, R., Olmi, L., Testi, L., Walmsley, C.M., and Hofner, P.: 2005, *A&A* in press.

Chini, R., Hoffmeister, V., Kimeswenger, S., Nielbock, M., Nürnberger, D., Schmidtobreick, L., Sterzik, M.: 2004, *Nature* **429**, 155.

Cohen, R.J., Brebner, G.C., Gasiprong, N., Hutawarakorn, B., Graham, M., and Meaburn, J.: 2004, *in Star Formation at High Angular Resolution, ASP Conference Series, IAU Symposium*, **221**, in press.

De Buizer, J.M.: 2003, *MNRAS* **341**, 277.

Edris, K.A., Fuller, G.A., Cohen, R.J., Etoka, S.: 2005, *A&A* in press.

Fiebig, D., Duschl, W.J., Menten, K.M., and Tscharnuter, W.M.: 1996, *A&A* **310**, 199.

Furuya, R.S., Cesaroni, R., Codella, C., Testi, L., Bachiller, R., and Tafalla, M.: 2002, *A&A* **390**, L1.

Garay, G. and Lizano, S.: 1999, *PASP* **111**, 1049.

Gibb, A.G., Hoare, M.G., Mundy, L.G., and Wyrowski, F.: 2004, *in Star Formation at High Angular Resolution, ASP Conference Series, IAU Symposium*, **221**, in press.

Greenhill, L.J., Reid, M.J., Chandler, C.J., Diamond, P.J., and Elitzur, M.: 2004, *in Star Formation at High Angular Resolution, ASP Conference Series, IAU Symposium*, **221**, 155.

Guilloteau, S., Dutrey, A., and Simon, M.: 1999, *A&A* **348**, 570.

Hoare, M.G., Glindemann, A., and Richichi, A.: 1996, *in The Role of Dust in the Formation of Stars, ESO Proc., eds. H.U. Kaüfl and R. Siebenmorgen*, 35.

Hofner, P., Cesaroni, R., Rodríguez, L.F., and Martí, J.: 1999b, *A&A* **345**, L43.

Keto, E.R., Ho, P.T.P., and Haschick, A.D.: 1988, *ApJ* **324**, 920.

Minier, V., Booth, R.S., and Conway, J.E.: 1998, *A&A* **336**, L5.

Minier, V., Booth, R.S., and Conway, J.E.: 2000, *A&A* **362**, 1093.

Moscadelli, L., Cesaroni, R., and Rioja, M.J.: 2000, *A&A* **360**, 663.

Norris, R.P., Byleveld, S.E., Diamond, P.J., Ellingsen, S.P., Ferris, R.H., Gough, R.G., Kesteven, M.J., McCulloch, P.M., Phillips, C.J., Reynolds, J.E., Tzioumis, A.K., Takahashi, Y., Troup, E.R., and Wellington, K.J.: 1998, *ApJ* **508**, 275.

Olmi, L., Cesaroni, R., Hofner, P., Kurtz, S., Churchwell, E., and Walmsley, C.M.: 2003, *A&A* **407**, 225.

Palla, F. and Stahler, S.W.: 1993, *ApJ* **418**, 414.

Pestalozzi, M.R., Elitzur, M., Conway, J.E., Booth, R.S.: 1994, *ApJ* **603**, L113.

Phillips, C.J., Norris, R.P., Ellingsen, S.P., and McCulloch, P.M.: 1998, *MNRAS* **300**, 1131.

Sandell, G., Wright, M., and Forster, J.R.: 2003, *ApJ* **590**, L45.

Shepherd, D.S. and Kurtz, S.: 1999, *ApJ* **523**, 690.

Shepherd, D.S., Yu, K.C., Bally, J., and Testi, L.: 2000, *ApJ* **535**, 833.

Shepherd, D.S., Claussen, M.J., and Kurtz, S.: 1999, *ApJ* **523**, 690.

Shu, F.H., Adams, F.C., and Lizano, S.: 1987, *ARA&A* **25**, 23.

Simon, M., Dutrey, A., and Guilloteau, S.: 2000, *ApJ* **545**, 1034.

Tan, J.C. and McKee, C.F.: 2002, *Hot Star Workshop III: The Earliest Stages of Massive Star Birth. ASP Conference Proceedings*, **267**, 267.

Tofani, G., Felli, M., Taylor, G.B., and Hunter, T.R.: 1995, *A&AS* **112**, 299.

Torrelles, J.M., Gómez, J.F., Rodríguez, L.F., Curiel, S., Ho, P.T.P., and Garay, G.: 1996, *ApJ* **457**, L107.

Yorke, H.W.: 2004, *Star Formation at High Angular Resolution, ASP Conference Series, IAU Symposium*, **221**, in press.

Yorke, H.W. and Sonnhalter, C.: 2002, *ApJ* **569**, 846.

Zhang, Q., Ho, P.T.P., and Ohashi, N.: 1998a, *ApJ* **494**, 636.

Zhang, Q., Hunter, T.R., and Sridharan, T.K.: 1998b, *ApJ* **505**, L151.

Zhang, Q., Hunter, T.R., Sridharan, T.K., and Ho, P.T.P.: 2002, *ApJ* **566**, 982.

# EMBEDDED CLUSTERS

Elizabeth A. Lada
*Astronomy Department, University of Florida, Gainesville, Fl 32608*
lada@astro.ufl.edu

**Abstract**

Star formation is a continuous ongoing process occurring over the lifetime of our Galaxy and the universe. During the last two decades we have made remarkable progress toward understanding star formation due to advances in observational technology especially at infrared and millimeter wavelengths which allow direct observation of the sites of star birth. Such observations suggest that embedded clusters may be the fundamental units of star formation in molecular clouds. Low star formation efficiency and rapid gas dispersal make these clusters disperse to provide the field star population. Consequently embedded clusters provide important laboratories for investigating fundamental issues of star formation within our Galaxy.

## Introduction

Stars are the most basic objects in the universe and how stars and subsequently their planets form is one of the most fundamental unsolved mysteries of astrophysics. Unraveling the process of stellar birth is not only of fundamental importance for understanding the origins of stars themselves, but also the origin and evolution of planets, life and even galaxies. For example, stellar formation and evolution are the engines that drive galaxy evolution and any progress toward understanding galaxy formation and evolution will ultimately be tied a our theoretical understanding of the basic physical processes which govern the star formation process.

Our observational understanding of star formation has undergone a radical change over the past 15 years. Two decades ago, we worked toward unraveling the processes of star formation by concentrating our studies on regions and objects that were thought to be the paradigm for the formation of sun like stars, mainly isolated, low mass molecular cores forming only one or a few low mass stars. Much progress was made in this endeavor both observationally and theoretically and has provided an important foundation for understanding the star forming process. However our focus has dramatically changed. This is mainly due to discovery of large numbers of embedded clusters by the use of infrared array cameras and the subsequent realization that these young clusters are responsible for a significant fraction of all star formation currently occurring in the Galaxy. Indeed, embedded clusters may be the fundamental units of star formation in giant molecular clouds.

## Importance of Embedded Clusters

Embedded clusters are best identified at near-infrared wavelengths, where the effects of extinction are minimized. Consequently, it was not until the late 1980s, when near-infrared imaging arrays were available for astronomical uses, that large numbers of embedded clusters were discovered and studied. Searches of the astronomical literature over this period in time by Lada & Lada (2003) and by Porras et. al (2003) have revealed that well over a hundred young embedded clusters have been found throughout the Galaxy. Most of these clusters were discovered via infrared surveys of regions having some signpost of star formation. For example the Hodapp (1994) K-band imaging survey of molecular outflows was particularly successful at finding embedded clusters. Surveys conducted using the data generated by the all sky near-infrared surveys such as 2MASS and DENIS will likely provide the most systematic and complete inventory of the local embedded cluster population in our Galaxy. Recent work by Bica et al. (2003) and Dutra et al. (2003) using the 2MASS data base has increased the number of known embedded clusters by more than 50%.

The discovery of such large numbers of embedded clusters emphasizes their importance for understanding the process of star formation in our Galaxy. However, it does not allow us to estimate the fraction of stars born within embedded clusters. This is best achieved from large scale surveys of molecular clouds. The first systematic attempt to obtain an inventory of high and low mass star formation in a single GMC was made by Lada et al (1991) who performed an extensive K-band near-infrared imaging survey of the central regions ($\sim$ 1 square degree) of the L1630 GMC in Orion. This survey produced the unexpected result that the vast majority (60-90%) of the star formation in that cloud occurred within a few rich cluster with little activity in the vast molecular cloud regions outside these clusters. Subsequent near-infrared surveys of L1630 as well as of other molecular clouds (Li, Evans & Lada 1997; Lada et al 1994; Carpenter, Snell & Schloerb 1995; Carpenter 2000) have yielded similar results with estimates of 50-100% of the clouds' embedded populations confined to embedded clusters.

While such near-infrared imaging observations are the most effective means of obtaining a census of the youngest stars in molecular clouds, they have limitations. Namely, near-infrared surveys can be contaminated by field stars in front of and behind the molecular clouds and can also suffer from background source confusion at very faint magnitudes. The general field star population consists mainly of late type dwarf and giants which are bright at near-infrared wavelengths. Given the reduced extinction at these wavelengths such stars can contribute significantly to star counts obtained at 2 microns even when they are observed through an intervening molecular cloud. The most common way to determine the contribution of field stars to a near-IR survey is to obtain star counts of control fields, large areas located nearby but off the cloud. By appropriately scaling the area of these control fields to match the area surveyed on the cloud, one can determine the membership of young stars in the cloud on a statistical basis. As one images to fainter and fainter levels the number of background stars increases and the statistical uncertainty in the star density within the molecular cloud increases. Thus for faint low mass stars it is difficult to place stringent limits on their numbers due to the "noise" in the expected background distribution.

This uncertainty can be significantly reduced, if we know that a star is likely associated with the star forming region. Near-infrared spectroscopy can distinguish young stellar objects from field stars. However, given the number of stars observed toward GMCs, both associated with the cloud and in the field, obtaining complete spectroscopic samples is extremely difficult in a reasonable amount of observing time. The development of near-infrared multi-object spectrometers, such as FLAMINGOS (Elston 2003) is essential for progress in this area. Multi-color near-infrared photometry can also be used to associate young stars with their parent clouds. Since young stars are surrounded by large amounts of dust, they exhibit excess emission above their photospheres at infrared wavelengths, starting at 2 microns. Detecting near-infrared excess emission through multi-color photometry, can therefore significantly reduce the background of stars in comparison to single wavelength star count observations.

The FLAMINGOS Star Formation Survey has recently obtained sensitive (i.e. $K < 17.5$) J,H and K images of $\sim$ 16 square degrees of sky in four nearby GMCs, Perseus, Orion, Monocerous and the Rosette. In addition, $\sim$ 1000 FLAMINGOS near-infrared spectra have already been obtained for selected stars in these clouds. Figure 1 presents our initial imaging results for the Rosette Molecular Cloud. From a JHK color analysis of the distribution of near-infrared excess sources, 13 young embedded clusters have been identified within the molecular cloud (Roman 2005; Roman et al. 2005). This is roughly a factor of 2 increase in the number of clusters previously found in this region (i.e. Phelps & Lada 1999). Furthermore the distribution of near-infrared excess sources indicates that the star formation in the cloud is highly clustered with more than 75% of the star formation in these clusters covering $\sim$ 20%

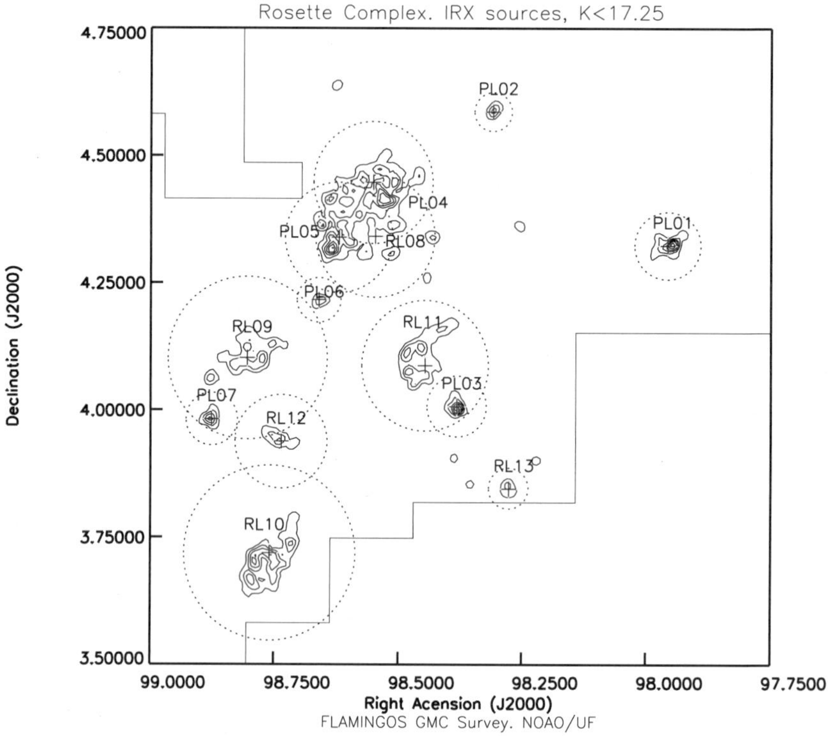


*Figure 1.* The Distribution of near-infrared excess sources in the Rosette Molecular Cloud is shown. The contours represent the density of excess sources having 10th nearest neighbor (10NN) densities greater than 1.5 arcmin$^{-2}$. The lowest level is 5 times the mean background density and subsequent levels are multiples of this level. The dashed circles represent an area with a radii equal to twice the cluster radii. A significant number of excess sources having lower 10NN densities are located within these "halos". The solid lines show the boundaries of the survey area in this figure.

of the molecular cloud area. These findings are consistent with the studies of other molecular clouds as discussed above and strengthen the conclusion that embedded clusters may indeed be the fundamental units of star formation in GMCs and hence the Galaxy.

## Global Properties of Embedded Clusters

Since clusters now appear to be the fundamental unit of star formation rather than individual stars, it is important to study the integrated or ensemble properties of young clusters in order to understand how star formation proceeds on the Galactic scale. Determining quantities such as the embedded cluster mass function, cluster birthrate and global star formation efficiency allow us to investigate the origin and evolution of clusters and hence lay a foundation for understanding galaxy evolution.

## Embedded Cluster Mass Function

Lada & Lada (2003) derived an embedded cluster mass distribution function (ECMDF) using their embedded cluster catalog compiled from the literature and including clusters having more than 35 stars and at a distance less than 2.4 kpc (Figure 2). Their derived mass distribution function reveals two potentially significant features. First, the function is relatively flat over a range spanning at least an order of magnitude in cluster mass (i.e., $50 \leq M_{ec} \leq 1000$ $M_\odot$). This suggests that massive clusters($\sim$ 1000 $M_\odot$), while rare, make up a significant fraction of the total stellar mass, roughly the same as for the more numerous lower mass clusters ($\sim$ 50-100 $M_\odot$) . Furthermore, more than 90% of the stars in clusters are found in clusters with masses in excess of 50 $M_\odot$ corresponding to populations in excess of 100 members. A similar result has been reported by Porras et al. (2003), who find that although the majority of clusters in their catalog are small with fewer than 30 members, most stars ($\sim$ 80%) form in clusters having $>$ 100 members. The flat mass distribution corresponds to an embedded cluster mass spectrum (dN/d$M_{ec}$) with a spectral index of -2 over the same range. This value is quite similar to the spectral index (-1.7) typically derived for the mass spectrum of dense molecular cloud cores (e.g. Lada, Bally & Stark. 1991), possibly suggesting that a uniform star formation efficiency characterizes most cluster forming dense cores. The index for the mass spectrum of embedded clusters is also essentially the same as that (-1.5 to -2) of classical open clusters (e.g., van den Berg & Lafontaine 1984; Elmegreen & Efremov 1997).

The second important feature in the ECMDF is the apparent drop off in the lowest mass bin ($\sim$ 20-50 $M_\odot$). Given that the Lada & Lada cluster catalog only included clusters with more than 35 stars, it is likely that it is considerably more incomplete for clusters in the 20 to 50 $M_\odot$ range than for the higher mass

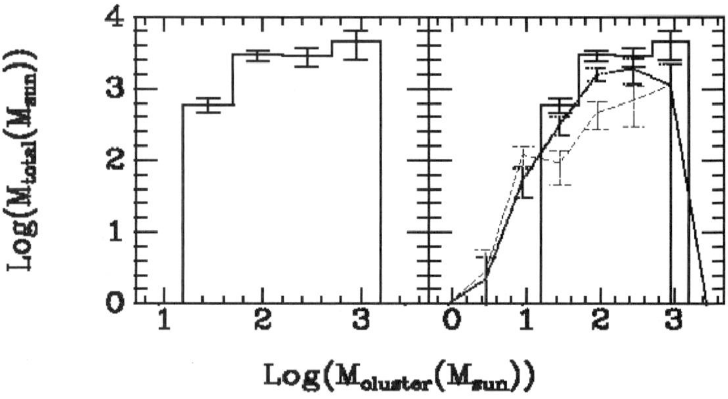


*Figure 2.* The embedded cluster mass distribution function (ECMDF) from Lada & Lada (2003). On the left the ECMDF is plotted for clusters with N>35 stars and located within 2KPC from the entire LL catalog. The right panel compares the ECMDF with those derived for two cluster subsamples which are believed to be more complete at the lowest masses. The dotted line is the Hodapp outflow sample and the dashed line the sample of all known embedded clusters within 500 pc of the Sun. The ECMDFs all appear to decline below 50 $M_\odot$.

clusters. To test the significance of this fall off to low cluster masses they also derived the mass distribution functions for subsets of clusters drawn from 3 homogeneous samples and from this analysis concluded that the drop off in the ECMDF at masses less than 50$M_\odot$ is significant. Apparently, there appears to be a characteristic cluster mass (50 $M_\odot$) above which the bulk of the star forming activity in clusters is occurring.

## Embedded Cluster Birthrate and Survival

Early estimates of the embedded cluster birthrate, based primarily on the number of clusters in the Orion cloud complex, found the rate to be extremely high compared to the birthrate of classical open clusters suggesting that only a small fraction of embedded clusters survived emergence from molecular clouds to become classical open clusters (Lada & Lada 1991). Using their more extensive embedded cluster catalog, Lada & Lada (2003) estimate a lower limit of the formation rate for clusters within 2kpc to be between 2-4 clusters $Myr^{-1}$ $kpc^{-2}$ for assumed average embedded cluster ages of 2 and 1 Myrs, respectively. Although this rate is a lower limit, it is a factor of 8–17 times that (0.25 $Myr^{-1}$ $kpc^{-2}$) estimated for classical open clusters by Elmegreen & Clemens (1985) and 5-9 times that (0.45 $Myr^{-1}$ $kpc^{-2}$) estimated by Battinelli & Capuzzo-Dolcetta (1991) for a more complete open cluster sample within 2 kpc of the Sun. This difference in birthrates between

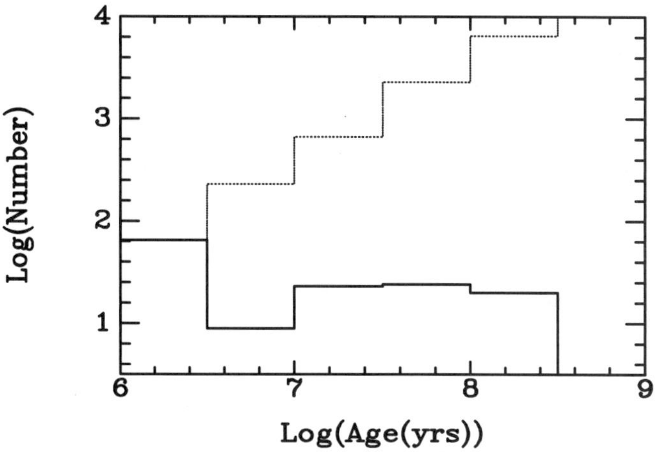


*Figure 3.* Observed frequency distribution of ages for open and embedded clusters within 2 Kpc of the Sun (solid line) compared to that (dotted line) predicted for a constant rate of star formation adjusted for cluster luminosity evolution. All embedded clusters fall into the first bin. The large discrepancy between the predicted and observed numbers indicates a high infant mortality rate for protoclusters. Taken from Lada & Lada (2003)

embedded and open clusters represents an enormous discrepancy and is of fundamental significance for understanding cluster formation and evolution.

Figure 3 shows the observed distribution of ages of all known clusters both embedded and open within 2 kpc produced from the combined catalogs of Lada & Lada (2003) and Battinelli & Capuzzo-Dolcetta (1991). Embedded clusters populate the lowest age bin. The number of clusters is found to be roughly constant as a function of age for at least 100 Myr. Also plotted in this figure is the predicted cluster age distribution, assuming a constant rate of star formation and adjusting for cluster luminosity evolution. There is a large and increasing discrepancy between the expected and observed numbers. These distributions clearly confirm earlier speculations that the vast majority of embedded clusters do not survive emergence from molecular clouds as identifiable systems for periods even as long as 10 Myr. *Figure 3 suggests an extremely high infant mortality rate for clusters.*

Less than $\sim$ 4–7% of the clusters formed in molecular clouds are able to reach ages beyond 100 Myr in the solar neighborhood, less than 10% survive longer than 10 Myr. Indeed, most clusters may dissolve well before they reach an age of 10 Myr. It is likely that only the most massive clusters are candidates for long term survival. Roughly 7% of embedded clusters in the Lada & Lada (2003) catalog have masses in excess of 500 $M_\odot$, and this likely represents a lower limit to the mass of an embedded cluster that can evolve to a Pleiades-like system. Moreover, Figure 3 also indicates that the disruption rate for bound clusters between 10–100 Myrs of age is significant, probably due to encounters with GMCs. Many of the observed open clusters in this age range may also not be presently bound (Battinelli & Capuzzo-Dolcetta 1991).

The observed low star formation efficiencies (SFEs) for embedded clusters (i.e. $<30\%$; Lada 1992; Lada & Lada 2003) can account for the high infant mortality rate of embedded clusters. Such SFEs indicate that the gravitational binding energy in an embedded cluster system is mostly provided by the parental molecular gas and therefore the fate of the cluster depends sensitively on the timescale for gas removal. If the gas is removed on timescales shorter than a dynamical time, then a cluster will remain bound only if the SFE $>50\%$ (Wilking and Lada 1983). Consequently, although most stars form in embedded clusters, the stars in these systems dynamically evolve to become members of unbound associations and only the most massive embedded clusters emerge as bound entities. However, bound classical clusters form at a sufficiently high rate that on average, that each OB association or GMC complex probably produces on such bound system (Elmegreen & Clemens 1985) accounting for about 10% of all stars formed within the Galaxy (Roberts 1957; Adams & Myers 2001).

## Role of Embedded Clusters in Understanding Star and Planet Formation

If embedded clusters are indeed the fundamental units of star formation in the Galaxy, then the properties of the stars within such clusters should determine the properties of the field stars in the Galaxy as a whole. Embedded clusters individually and collectively contain statistically significant numbers (hundreds) of young stellar objects (YSOs), providing a meaningful sampling of essentially the entire stellar mass function. Clusters can be characterized by the mean age of all its members, which statistically is a reliable indicator of age, and a sample of clusters can be observed which span a significantly wider range of age than is typical of stars formed in any individual star-forming region. Consequently, embedded clusters provide important laboratories to determine the initial mass function (IMF), binarity, origin of brown dwarfs, the origin and evolution of circumstellar disks and the likelihood of planet formation. There is not enough space to discuss each of these issues here and the reader is referred to the review by Lada & Lada (2003). Here we will discuss the role that embedded clusters play in determining the frequency and evolution of circumstellar disks and the implications for planet formation.

## Protoplanetary, Circumstellar Disks

Knowledge of the frequency distribution of circumstellar disks in clusters is key to understanding the processes involved in the formation and evolution of circumstellar disks and any planetary systems. For example, in the youngest embedded clusters, determination of the disk frequency is a measure of the probability of disk formation around newly formed stars. In addition, the investigation of the variation of disk frequency with cluster age, provides a direct measure of the lifetimes of circumstellar disks and hence the duration for planet building.

Recently, Haisch, Lada, & Lada (2001) performed the first systematic and homogeneous observational survey for circumstellar disks in young clusters. They used JHKL imaging observations to determine the circumstellar disk frequencies in six clusters whose ages ranged from 0.2 - 30 Myr. These observations reveal that clusters are characterized by a very high initial disk frequency ($\geq 80\%$) which then sharply decreases with cluster age (Figure 4). Half the disks in a cluster population are lost in only about 3 Myr, and the timescale for essentially all the stars to lose their disks appears to be about 6 Myrs (Haisch, Lada, & Lada 2001; hereafter HLL01). Such short disk lifetimes, combined with the fact that many of the extrasolar planets discovered to date are in very close proximity to their central stars as a result of migration (e.g., Marcy & Butler 1999), would appear to support models which can accommodate rapid planetary formation (e.g., Boss 1998, 2000). Near-infrared observations, however, only probe the *inner* disk ($\leq$ 0.25 AU) via reflected light or hot (900K) dust emission. The lifetime for the *outer* disk regions ($\geq$ 1 AU), which contain the bulk of the mass available to form planets, may be different from that for the inner disk. For example, if disks evolve large inner holes as they age, then circumstellar disk lifetimes determined from NIR observations, which are sensitive to the inner portions of the disk, could be much shorter than those derived from observations of the outer disk. If the outer disk is not coupled to the inner disk and its lifetime is longer than that which we measure for the inner disk, models which predict a longer giant planet formation timescale could be accommodated (Thi et al. 2001; Lissauer 2001)

Millimeter wavelength observations provide direct probes of the masses and lifetimes of circumstellar disks in regions of the disks which are of interest for planet formation. Such observations trace the emission from cool dust which traces the bulk mass in the disk systems (Beckwith & Sargent 1993; Beckwith 1999). The unprecedented sensitivity and spatial resolution of existing large, millimeter-wave telescopes and sensitive bolometers allows the detection of thermal emission from circumstellar disks with total masses equal to that of Jupiter (or less under ideal observing conditions). Hence, it is only at millimeter wavelengths that the masses and planet forming capabilities of the

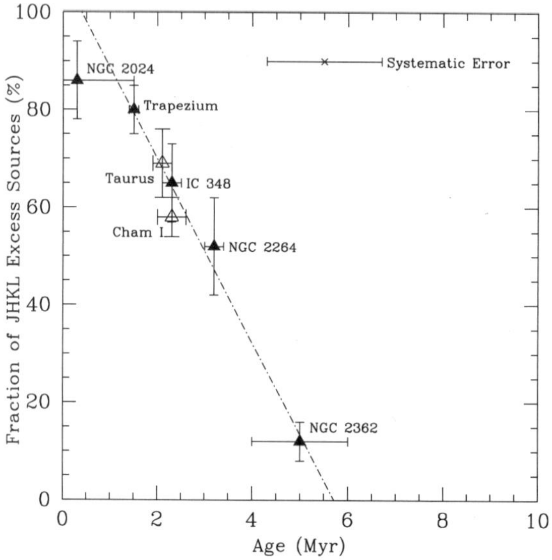


*Figure 4.* Disk Fraction as a function of cluster age for a sample of young cluster with consistently determined mean ages. The disk fraction is initially very high, but then rapidly drops with cluster age suggesting maximum disk lifetimes of less than 6 Myrs in young clusters (Haisch et al. 2001)

putative disk systems can really be assessed. To date, the most comprehensive millimeter continuum surveys for circumstellar disks have been conducted at 1.3 millimeters in the closest ($<$ 150pc) star forming regions such as Taurus (Beckwith et al. 1990; Osterloh & Beckwith 1995) and $\rho$ Oph (André & Montmerle 1994). These studies have shown that a large fraction of stars in these nearby regions have disks massive enough to form planets. Only a few attempts have been made to survey young rich clusters, other than $\rho$ Oph, at millimeter wavelengths with sensitivities sufficient to detect circumstellar disks capable of forming planets. For example the very young Trapezium cluster has been surveyed at $\sim$ 3 mm with both the BIMA (Mundy, Looney, & Lada 1995) and OVRO (Bally et al. 1998) interferometers. Neither study detected any massive disks greater than 0.015 $M_\odot$. IRAM interferometer continuum observations (Lada et al. 2000; Lada 1998)) obtained at 1.3 and 2.7 mm of this cluster reveal emission from $\sim$ 10% of the NIR sources in the fields observed. The masses for these circumstellar disks are estimated to be only $\sim$ 0.01 $M_\odot$. In addition, Carpenter (2002) surveyed 95 stars in the 2.3 Myr old, IC 348 cluster in 3 mm continuum using the OVRO interferometer. Again no massive disks were detected. Taken at face value, these results could imply that it is very difficult to form massive, planet forming disks in clusters or alternatively that disks are quickly destroyed in such environments.

Lada & Haisch (2005) have carried out a sensitive, systematic 1.3 mm continuum survey of four embedded clusters, NGC 1333, NGC 2071, NGC 2068 and IC 348. They detected $\sim$ 35% of the sources in the NGC 1333 and NGC 2071 clusters and $\sim$ 14% of the sources surveyed in NGC 2068 and IC 348. The relatively high occurrence of massive outer disks in the NGC 1333 and NGC 2071 clusters implies that these massive disks are common in young clusters. Comparison of these mm observations with near-IR observations of the same clusters reveal that the variation in the fraction of detected millimeter sources from cluster to cluster is similar to the variation in the fraction of near-IR excess sources for these clusters. This suggests that the inner disk and the outer disk are coupled and indicates that the decrease in the outer disk fraction is a result of evolution. Further, the very low occurrence of massive disks in the IC 348 cluster for which there exists good age estimates (Herbig 1998) strongly suggests that the outer disks in this cluster dissipate in less than 3 million years.

The short disk evolution timescales ($<$ 3–6 Myr) suggested by both the millimeter and near-infrared observations discussed above, place important constraints on the timescale for building gas giant planets in cluster environments. Typical timescales derived from the accepted model of planet formation, the core accretion model are roughly 10 Myrs or greater depending on the mass accretion rate of solid material, the disk surface density of solid material and atmospheric dust abundance (Bodenheimer, Hubickyj, & Lissauer 2002; Lissauer 2001). It may be very difficult to both build and then migrate planets (Lin, Bodenheimer, & Richardson 1996) on such short timescales with standard core accretion models, if the gas depletion time scale is as short. Indeed, evidence of short gas dissipation times may be indicated from studies of CO fundamental emission in the inner disks of young low mass stars (Najita 2003). The disk lifetimes measured would be sufficient for giant planets to form via models of Boss (1998, 2000) which invoke gravitational instabilities within relatively massive ($M_{disk} \sim 0.1\ M_\odot$ within a radius of 20 AU) protoplanetary disks to form giant planets. These models can produce giant planets on a much shorter ($\sim 10^3$ yr) timescale. However, more theoretical investigation is required to determine whether such models are really viable (Boss 2000).

## Acknowledgments

I am very grateful to Nanda Kumar and the organizing committee for kindly inviting me to participate at this interesting meeting. I thank Carlos Roman for help with the Rosette figure. I acknowledge support from a Presidential Early

Career Award for Scientist and Engineers (NSF AST 97-33367) and from NSF Grant NSF AST 02-04976 to the University of Florida.

## References

Adams, F. C. & Myers, P. (2001) *ApJ* 533:744

André, P. & Montmerle, T. (1994) *ApJ* 420:837

Bally, J., Testi, L., Sargent, A. & Carlstrom, J. (1998) *AJ* 116:854

Battinelli, P. & Capuzzo-Dolcetta (1991) *MNRAS* 249:76

Beckwith, S. V. W. & Sargent, A. I. (1993) in *Protostars and Planets III, edited by E. H. Levy and J. I. Lunine, (Tucson: Univ. Arizona Press)* 521

Beckwith, S. V. W., Sargent, A. I., Chini, R. S., & Guesten, R. (1990) *AJ* 99:924

Beckwith, S. V. W. (1999) in *The Origin of Stars and Planetary Systems, edited by C. J. Lada and N. Kylafis (Kluwer, Dordrecht)* 579

Bica, E. Dutra, C. M. Soares, J., Barbuy, B. (2003). *A&A* 404:223

Bodenheimer, P., Hubickyj, O., & Lissauer, J. J. (2000). *Icarus* 143:2

Boss, A. P. (1998). *ApJ* 503:923

Boss, A. P. (2000). *ApJ* 536:L101

Carpenter, J. (2002). *AJ* 124:1593

Carpenter, J. M., Heyer, M. H. & Snell, R. L. (2000). *ApJS*, 130, 381

Carpenter, J. M. (2000). *AJ* 120:3139

Carpenter, J. M., Snell, R. L. & Schleorb, F. P. (1995). *ApJ* 450:201

Dutra, C. M., Bica, E., Soares, J., Barbuy, B. (2003). *A&A* 400:533

Elmegreen, B. G & Clemens, C. (1985). *ApJ*, 294, 523

Elmegreen, B. G. & Efremov Y. N. (1997). *ApJ* 480:235

Haisch, K. E., Lada, E. A., & Lada, C. J. (2001). *ApJ* 553:L153

Herbig, G. (1998). *ApJ* 497:736

Hodapp, K. (1994). *ApJS* 94:615

Lada, C. J. & Wilking, B. A. (1984). *ApJ* 287:610

Lada, C. J. & Lada, E. A. (1991). in *The Formation and Evolution of Star Clusters, ed. K. Janes, (San Francisco: Astronomical Society of the Pacific)* 3

Lada, C. J., Lada E. A., Clemens, D. P. & Bally, J. (1994) *ApJ* 429:694

Lada, C. J. & Lada E. A. (2003). *ARA&A* 41:57

Lada, E. A., Bally, J & Stark, (1991). *ApJ* 368:432

Lada, E. A., DePoy, D. L., Evans, J. H. & Gatley, I. (1991). *ApJ* 371:171

Lada, E. A. (1992) *ApJ* 393:L25

Lada, E. A. (1998) in *The Origin of Stars and Planetary Systems, eds. C. J. Lada & N. D. Kylafis, (Dordrecht: Kluwer)* 441

Lada, E. A. & Haisch, K (2005). *ApJ* submitted

Lada, E. A., Dutrey, A. Guilloteau, S. Haisch, K. & Mundy, L. (2003) in prep

Li, W., Evans, N. J. & Lada, E. A. (1997) *ApJ* 488:277

Lin, D. N. C., Bodenheimer, P., & Richardson, D. C. (1996). *Nature* 380:606

Lissauer, J. J. (2001) *Nature* 409:23

Marcy, G. W. & Butler, R. P. (1999) in *The Origin of Stars and Planetary Systems. ed. Charles J. Lada & Nikolaos D. Kylafis. (Kluwer, Dordrecht)* 681

Mundy, L.G., Looney, L.W. & Lada, E. A. (1995) *ApJ* 452:L137

Najita, J. *AAS*

Osterloh, M., & Beckwith, S. V. W. (1995) *ApJ* 439:288

Phelps, R. & Lada, E. A. (1997) *ApJ* 477:176

Porras, A. Christopher, M., Allen, L., Di Francesco, J., Megeath, S. T. & Myers, P. C. (2003) *AJ* 126:191

Roberts, M. S. (1957) *PASP* 69:59

Roman, C. (2005) *PHD Thesis, University of Flordia*

Thi, W. F., Blacke, G. A., von Dishoeck, E. F., van Zadelhoff, G. H., Horn, J. M. M., et al. (2001) *Nature* 409:60

van den Berg, S & Lafontaine, A. (1984) *AJ* 89:1822

Elizabeth Lada and Teresa Lago. Zoltan Balog taking a photo at the back.

# MASSIVE PROTOSTARS AND SMALL PROTOCLUSTERS

M. S. N. Kumar
*Centro de Astrofísica da Universidade do Porto, Rua das Estrelas, 7150-462 Porto, Portugal*
nanda@astro.up.pt

**Abstract**

Recent investigations of massive protostellar candidates in the near-infrared (NIR) wavelengths reveal several $2\mu$m visible embedded clusters, NIR counterparts to massive star precursors and $H_2$ emission in the shape of rings and flared disks around the massive protostars. These observations have shown that in an embedded cluster, the massive stars likely form after the first generation of low mass stars are born. The $H_2$ emission in the shape of rings and flared disks represent huge tori of molecular material around massive protostars. These two results in conjunction are consistent with the theory of massive star formation by continuing accretion. The clusters associated with very young precursors to massive stars may represent an evolutionary stage prior to dynamical relaxation. Detailed studies of two clusters reveal flattened, ring shaped morphology both in stellar content and molecular gas. These ring clusters are consistent with the prediction of a theory to form small clusters through fragmentation of magnetically subcritical cloud to multiple magnetically supercritical cores.

## Massive Protostars

Our present understanding of the physics of star formation is mostly derived from the observational studies of low and intermediate mass protostars. Massive protostars have remained less understood objects due to their short formation time scales, relatively larger distances and higher extinction of the dense regions in which they are born. In recent years, much effort has gone into defining the properties of massive protostars and identifying them (see the reviews by Beuther & Shepherd and by Cesaroni in this book). Since massive stars are expected to be born in the densest star-forming cores with large amounts of visual extinction, observational studies of massive protostars have been traditionally concentrated in the radio and far-infrared wavelengths. Recently, a variety of observational studies of the high mass protostellar objects (HMPOs) in the near-infrared wavelengths has lead us to a common ground to investigate the birth of massive stars and star clusters. In particular, the increased spatial resolution provided by NIR wavelengths has revealed details that were otherwise not seen even with high angular resolution radio observations. In the following I shall present evidence for i)clustering (sec. 2) ii)near-infrared counterparts (sec. 3) and iii) $H_2$ emission in the form of rings and flared disks (sec. 4) around HMPOs. Sec. 5 will discuss the scenario of massive star formation consequent to the above observational findings. Finally, in Sec. 6 I shall discuss the peculiarities of some (proto)clusters associated with HMPOs.

## Clustering around candidate precursors to massive stars

It is known from studies of stellar clusters that stars of spectral type earlier than B7 are invariably associated with groups or clusters of lower mass stars (Hillenbrand 1995, Testi et al. 1999). However, it is not clear if the massive stars and the associated lower mass stars were born simultaneously, at different epochs, or if these associations are a consequence of some dynamical process in the star forming cloud.

In recent years systematic efforts have been made to identify the earliest phases of massive stars resulting in lists of candidate precursors to massive stars (Palla et al. 1991; Molinari et al. 1996; Sridharan et al. 2001). Since massive stars likely form in clusters, these lists of precursors to massive stars are also lists of stellar clusters that are still in the process of formation. Molinari et al. 1996 (hereafter MBCP96) targets have been separated into 'low' and 'high' categories of which the low sources are thought to be better candidates of massive protostars, while 'high' sources are likely to be more evolved, already in the ultra-compact HII region phase. Sridharan et al 2001 (hereafter Sri02) chose luminous IRAS point source targets that satisfied the Wood & Churchwell 1989 criteria for UCHII regions and simultaneously showed a lack of significant free-free radio (cm) continuum emission to a limit of 25mJy. The sample of targets derived from the combined lists of MBCP96 (163 targets; 80 low and 83 high) and Sri02 (69 targets) thus represent the best known candidate precursors to massive stars in the northern hemisphere. The $2\mu$m All Sky Survey (2MASS) database was used to search for clustering around the carefully selected sample of candidate precursors to massive stars. The goal of this search is two fold; a) to identify and study groups, associations or clusters around the selected sample of candidates, and b) to identify the near-infrared (NIR) counterparts to the HMPOs that are identified by their well defined peaks in the 1.2 mm continuum emission.

The near-infrared K-band at $2.2\mu$m suffers the least extinction of all the three J, H, & K-bands and therefore the photometric data in the K-band is the basis for detection of any groups, associations or clusters associated with the targets in the sample list. To detect such groups, stellar surface density maps were produced by binning the 2MASS point sources in the spatial dimension. This method has been previously applied in the studies of embedded clusters (see for ex: Lada & Lada 1995, Hillenbrand & Hartmann 1998 and Kumar et al. 2004). Using bins of 120" separated by 60" over an area of 600"×600" around each target, Nyquist sampled stellar surface density contour maps were produced. Clusters were detected as star count density enhancements above the mean background level within the sampled area. Contours of mode+2$\sigma$ and above were used to identify clusters/groups, where $\sigma$ and the mode value represents the average noise and the star count in the region.

*Table 1.* Cluster detection statistics for the 217 targets

| Source Type | Category | Number of targets | Definite clusters | Possible clusters | Total clusters | Detection rate |
|---|---|---|---|---|---|---|
| A | no cm contnm | 69 | 13 | 2 | 15 | 22% |
| B | MBCP96 low | 83 | 17 | 9 | 26 | 31% |
| C | MBCP96 high | 80 | 21 | 5 | 26 | 33% |
| | Total [common] | 217[15] | 47 [4] | 16 [0] | 63 [4] | 29% |

The number in brackets indicate the overlaps between MBFP96 and Sri02 sources

The detection statistics of clusters associated with the 217 unique targets are listed in Table. 1. As indicated, 54 clusters are detected out of 217 unique sources which implies a ~25% detection rate. It appears that targets without any significant radio continuum emission (type A) have the least association with clusters and the MBCP96 high sources (type C), which are very luminous targets associated with an UCHII region, have the highest association with clusters. If the target types A, B, C in Table 1 represent an evolutionary sequence of massive protostars, then there is a higher rate of cluster detection with more evolved sources as compared to younger sources. Using the JHK data from 2MASS, color-color and color-magnitude diagrams were constructed for each of the clusters detected. The number of stars in the cluster and the average extinction were estimated to each of these clusters. These data are used to compute an approximate mass for the cluster following the recipe of Lada & Lada (2003). These mass estimates serve for a relative comparision among the clusters and do not represent the actual mass of the observed cluster. This is because the 2MASS spatial resolution and photometric depth does not effectively sample the complete membership of the cluster owing to their large distances. However, these mass estimates allow us to construct an Embedded Cluster Mass Distribution Function (ECMDF) since the relative masses of the clusters are valid. Figure 1a shows such an ECMDF, where the solid line represents the ECMDF for the sample of 54 embedded clusters around candidate massive protostars and the dotted line shows the LL03 ECMDF for embedded clusters within 2 kpc from the Sun. It is evident from Fig.1a that the ECMDF of this sample conforms to the shape of LL03 ECMDF implying that most of the stellar mass is concentrated in few large clusters, although, there exist a large number of small clusters (see LL03 for a detailed discussion).

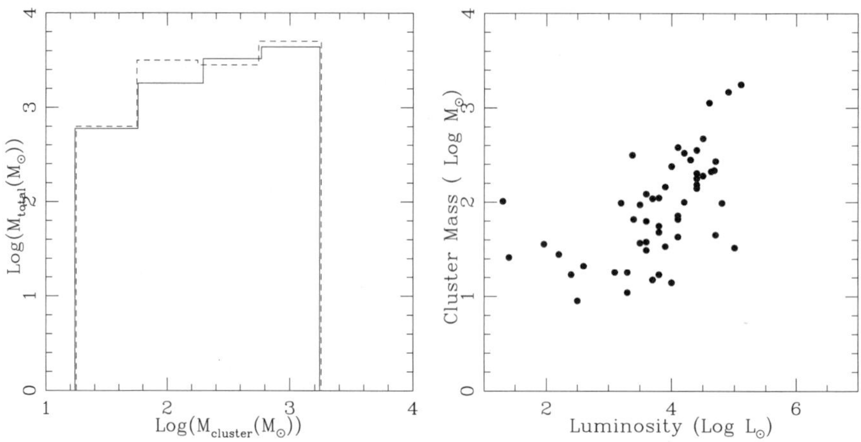


*Figure 1.* Embedded cluster mass distribution function. The dotted line represent the ECMDF from LL03 for the embedded clusters within 2Kpc distance. The solid line shows the ECMDF for the clusters from this work.

We can also plot the IR luminosity of the HMPO versus the mass of the associated embedded cluster as shown in Fig. 1b. As can be seen from the figure, there is a correlation between the mass of the associated embedded cluster and the luminosity of the target massive protostar implying that clustering may proceed as an increasing function of the luminosity of the HMPO.

## NIR counterparts to candidate massive protostars

Near-infrared color-color diagrams have been known to be powerful tools in identifying the nature of stars. The method was first analysed in detail and employed by Lada & Adams (1992) to identify the locus of HAeBe stars. Later it has been effectively used to identify low mass T-Tauri stars (Meyer et al. 1997), and even brown dwarfs (Muench et al. 2001). We extend this method to investigate the colors of the NIR counterparts of candidate HMPOs.

Beuther et al. (2002a) have fitted the central positions of observed peaks for each of the 69 HMPO sources that they mapped in the 1.2 mm continuum emission. These peaks of 1.2 mm emission are likely positions of embedded massive protostars. Similarly, dust continuum peaks for 30 IRAS sources are available from Molinari et al. 2000. We have searched for near-infrared counterparts within a radius of 5" to each of these 1.2 mm continuum peak positions and plotted them on a color-color diagram in Fig. 3. We first separated the dust

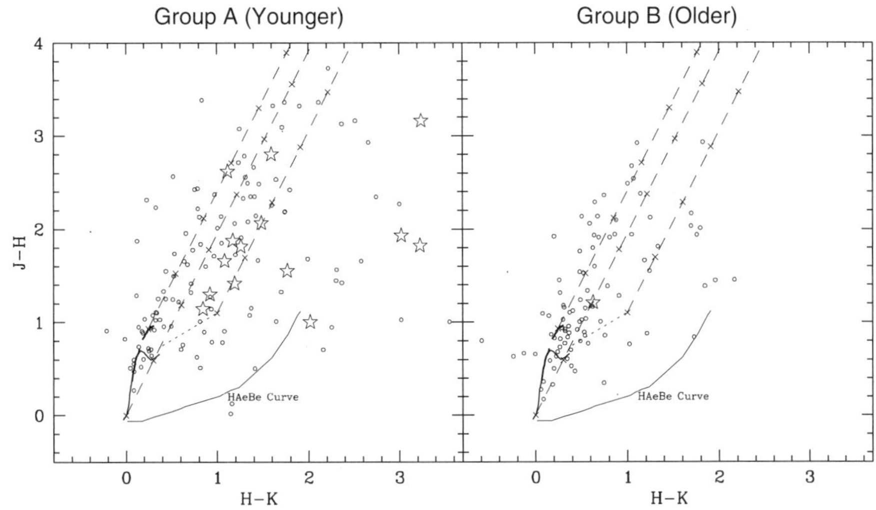


*Figure 2.* Color-color diagram of NIR counterparts to the 1.2 mm peaks of HMPOs. The solid curve represents the HAeBe curve by Lada & Adams (1992), thick-solid curve represents the main-sequence tracks, the dotted line shows the T-Tauri locus from Meyer et al.(1997) and the dashed lines represent the redenning vectors. The dots represent the point sources from GATOR PSC and star symbols represent the extended sources from the GATOR XSC.

continuum peaks into those that were and those that were not associated with radio free-free emission. The former were called 'younger' sources and the latter 'older' sources, respectively and are displayed in adjacent panels in Fig. 3. We searched for NIR counterparts using both the 2MASS Point Source Catalog (PSC) and Extended Source Catalog (XSC). The extended sources are shown as star symbols, and point sources are shown as points in Fig. 4. The main sequence dwarf & giant loci, the T-Tauri locus and the HAeBe locus are all plotted for reference. The younger sources display very red colors, are spread around the region of the main sequence and T-Tauri loci, and extend way beyond the limits of the HAeBe locus. In comparision, the older sources are more limited to the main-sequence region, some of them lying within the HAeBe locus. However, there is no clear pattern from these diagrams yet, that can be fit with a functional form. All the extended sources are associated only with younger targets with the exception of one that can be found on the plot of older targets. Many NIR-counterparts are visible only in the near-infrared H or just K-bands. However, good quality photometry of each NIR counterpart and color ratios such as H-K and K-L are needed to assess if there is a systematic pattern on the color-color diagrams that is followed by these massive protostars.

The extinction estimates from the 1.2 mm dust continuum maps of the candidate massive protostars (Beuther et al. 2002a) show high values of several hundred magnitudes of visual extinction suggesting that we may not see any $2\mu$m counterparts in these regions. However, the extinction estimates from millimeter studies are prone to some uncertainties, particularly due to uncertain values of assumed temperatures, the contamination of free-free emission from UCHII regions, and the assumed values of the dust emissivity index. In any case, the result described above has made it clear that despite the high extinction values measured from millimeter observations there are increasing numbers of near-infrared counterparts that precisely match the dust continuum peaks of luminous young stellar objects (see also, Kumar et al. 2002). If, indeed the high extinction estimates from the millimeter continuum emission measurements and the low values found from this near-infrared study, both are equally true, then it implies that the observed NIR counterparts are all situated at the periphery of dense cores that may contain large column densities. In any case, the visibility of a significant number of HMPOs in the NIR opens up the opportunity of using this wave-band to conduct high spatial resolution studies.

## $H_2$ emission around massive protostars

$H_2$ emission in the near-infrared K band is a good tracer of shocked and/or flouresced gas in star forming regions. In an effort to study the nature of Herbig-Haro (HH) flows associated with massive molecular outflows, Kumar et al. (2002) conducted a narrow-band imaging study of seven HMPOs in the $2.122\mu$m $H_2$ v=1–0 S(1) emission line. In the course of this investigation, they discovered for the first time, $H_2$ emission in the shape of rings and flared disks around NIR visible star-like counterparts of HMPOs. The sources IRAS 20293+3952 and IRAS 05358+3843 show well defined rings centered around $2\mu$m stars while flared disk structures are found around IRAS 07427-2400 and IRAS 20343+4129. The source IRAS 07427-2400 is one of the most luminous candidates of HMPOs among those studied by Kumar et al. (2002) and is also associated with an UCHII region. A detailed investigation of this source was pursued by using multiple tracers to find out the nature and origin of the flared disk shaped $H_2$ emission feature (Kumar et al. 2003b). The grey scale images in Fig. 3 show continuum subtracted narrow band image centered on the $H_2$ $2.122\mu$m line. The $2\mu$m star is subtracted out and appears as a whitish patch on this image. The position of this star and the UCHII region together is represented by a star symbol and is found to be centered on the flared disk-like feature. Fig. 3a shows contours of $350\mu$m dust continuum emission (heavy solid line), $C^{18}O$ emission (dotted line) and IRAS $12\mu$m emission(solid line). These tracers represent the dense gas and dust in the region and show an elliptical dense core. Fig. 3b show red and blue shifted contours of CO J=1–

0 emission representing a massive molecular outflow that is perpendicular to the axis of the observed $H_2$ emission. Thus the $H_2$ emission is arising in a disk/envelope which is extending up to 5" in radius which at the distance of 6.8kpc to the source, represents 50,000 AU. Given the presence of an UCHII region and a 2$\mu$m visible star, one would naturally expect that any $H_2$ emission close to the star is likely due to flourescent emission rather than shocked emission. However, detailed analysis of the ro-vibrational transitions of the v=1-0 and v=2-1 S(1) levels from $H_2$ using the near-infrared H and K band spectrum suggest that the emission is shocked rather than flouresced. Similarly, Beuther et al.(2004) present interferometric observations of the dust continuum emission and radio free-free emission overlaid on ring shaped $H_2$ emission feature of IRAS 20293+3952 which reveal a 2$\mu$m visible star identified as the massive protostar. The $H_2$ ring is not seen by any of the radio wavelength tracers.

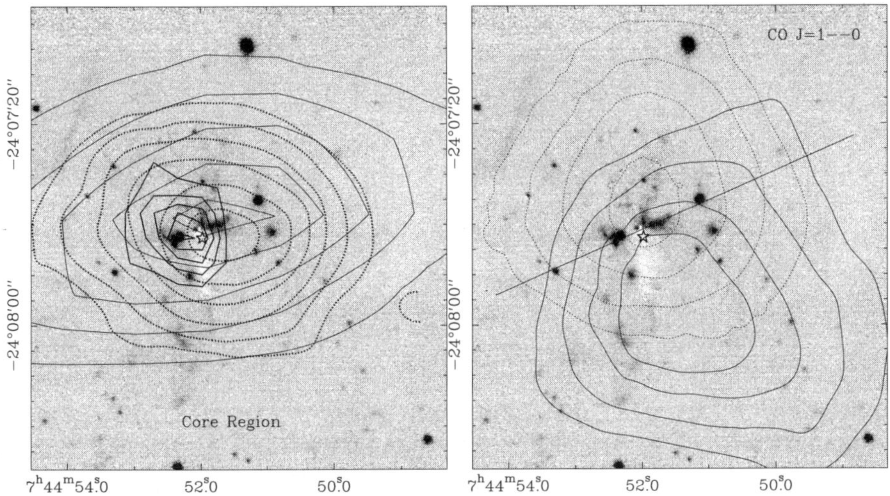


*Figure 3.* The flared disk structure around IRAS 07427-2400. The grey scale images show the continuum subtracted $H_2$ v=1–0 S(1) narrow-band image at 2.122$\mu$m. The block on the left shows 350$\mu m$ emission (thick contours), $C^{18}O$ emission (dotted contours) and IRAS 12$\mu$m emission (normal contours). The block on the right shows the red (dotted) and blue (solid) shifted CO J=1–0 emission tracing a massive outflow (from Shepherd & Churchwell 1996) arising from this source.

## Scenario for massive star formation

The observational evidences presented in the previous sections has given rise to a scenario that suggests accretion as the most prominent physical mechanism to form massive stars. The embedded clusters detected around a significant sample of HMPOs show near-infrared visible members which trace predominantly low mass population. Since the low mass population that constitutes the embedded clusters are already visible at $2\mu$m, the low mass stars must have crossed the Class 0-I phases of evolution and are most likely in Class II-III or PMS stages. In comparision, the massive protostars are passing through a phase like the Class 0 or I as evidenced, for example, by their association with massive outflows (Beuther et al. 2002b; Shepherd & Churchwell 1996). The observations indicate that in these clusters the low mass stars have already formed or are at the end of their formation phases while the high mass stars are still in their initial stages. Since low mass star formation lasts for approximately $10^6$ yrs while massive star formation, lasts perhaps for only $10^5$yrs, the observational finding of embedded clusters visible at $2\mu$m around candidate massive protostars shows that massive star formation in the clusters begins at least one-half to one Myrs after the first low mass stars have formed.

The NIR counterparts to HMPOs show characteristics of intermediate mass pre-main sequence stars or zero-age main sequence star based on their colors. These intermediate mass stars continue to be embedded in dense cores and are associated with rotating tori of gas and dust as evidenced by interferometric studies of several selected sources (see Cesaroni's review in this book). Therefore these observations suggest a scenario where intermediate mass young stellar objects are in the process of continuing accretion. Theoretical models indeed suggest such a mechanism for the formation of massive stars through continuing accretion on to intermediate mass stars (e.g. Meynet & Maeder 2000, Keto 2003; see also the article by Keto in this book). The age sequence of forming low mass stars first and then high mass stars in the evolution of a cluster, as envisaged above, are also in agreement with the theoretical calculations. The $H_2$ emission in the form of flared disks and rings, according to this scenario, can arise in the huge tori/envelopes around the HMPOs where matter is still accreting on to intermediate mass young stellar objects.

## Small proto-clusters

Embedded clusters where massive star formation is passing the protostellar phases, such as the clusters mentioned in Sec.2, are in a relatively early stage. As a cluster ages beyond a few crossing time scales, dynamical mixing becomes significant resulting in the loss of the original morphological structure with which the cluster was born. Thus, the clusters associated with isolated massive protostars probably represent some of the least dynamically relaxed states of observable embedded clusters. Recently, two clusters associated with HMPO sample were found to show a peculiar ring shaped morphology. These clusters associated with IRAS 22134+5834 and Mol148 have been investigated in detail for both stellar content and molecular gas. Fig.4 shows the K-band images of these clusters overlaid with contours of molecular emission.

Fig.4a shows Mol148 K-band image overlaid with $HCO^+$ 1–0 emission contours and Fig.4b shows a K-band image of IRAS 22134+5834 overlaid with $NH_3$ (J,K)(1,1) integrated emission contours. Although both clusters show ring shaped morphology, Mol148 displays this morphology prominently in the molecular gas content, while IRAS 22134+5834 reveals the same in its stellar content (see the front cover of this book for a JHK color composite of this cluster). The morphology is suggestive of a flattenned structure with a central cavity. The central regions that appear as a cavity have been searched for possible embedded sources by using the 1.2mm dust continuum emission and also at the 10$\mu$m mid-infrared band. In both cases, no embedded sources are found, and the 1.2mm peak is found to be situtated on one of the near-infrared visible stars of the ring. Recently, Spitzer Space Telescope images of the source Mol148 are available from the IRAC camera which is very sensitive to point sources in the mid-infrared up to 8$\mu$m. These images also justify the earlier finding that there are no embedded sources at the center of the ring.

Adams & Myers (2001) using several physical arguments have broadly defined embedded clusters with number of stars $100<N_\star<300$ as small clusters and $N_\star >300$ as large clusters. Following this definition, Kumar et al. ( 2004) concluded that IRAS 22134+5834 is a small protocluster and the same follows for Mol148 from recent deep NIR observations. Li & Nakamura ( 2002, hereafter LN02) presented calculations of the non-axisymmetric evolution of a magnetically subcritical molecular cloud under thin-disk approximation (flattenned cloud) fragmenting into multiple magnetically super-critical cores. Such calculations are believed to be fundamental to the formation of all varieties of star formation including singles, binaries, multiples and clusters. These authors predict that the supercritical cores resulting from fragmentation are arranged in a ring shape because the magnetic field tension prohibits the formation of a central singularity. The observations described above strongly support their predictions since the clusters Mol148 and IRAS 22134+5834 provides clear observational evidence for the presence of a ring in stellar content and molecular gas that encloses a central cavity. Indeed, the K-band image of IRAS 22134+5834 is in striking resemblance to the numerical simulations of LN02 for the case of perturbation with mode m=5 (Fig. 3c of LN02). If the above two clusters are representative of the small fraction of such clusters that are oriented face-on in the plane of the sky, they may account for a significant sample of such clusters in the sky when all orientations are considered. Therefore the scenario may well be valid for the formation of small protoclusters and the embedded clusters associated with HMPOs may truly contain important information about the time scales of dynamical relaxation.

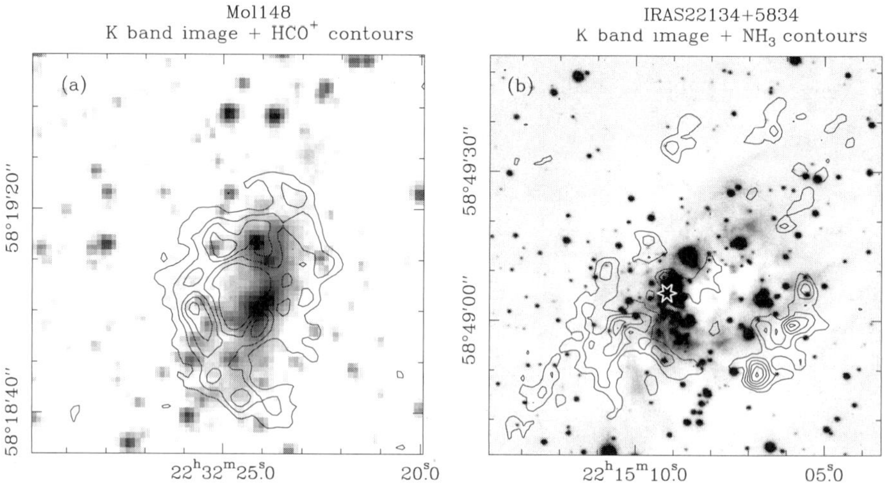


*Figure 4.* Clusters with ring shapes and central cavities. a) K-band image of Mol148 overlaid by $HCO^+$ J=1–0 integrated emission contours, b)K-band image of IRAS 22134+5834 overlaid by contours of $NH_3$(J,K)(1,1) integrated emission.

## Future Work

High resolution observations of HMPOs with present day interferometers such as IRAM Plateau de Bure or the Owens Valley Radio Observatory interferometer has provided ample evidence for the existence of dense cores, disks and outflows from HMPOs supporting the mechanism of accretion to form massive stars. However, the spatial resolution of $3''$-$5''$ offered by these facilities can not effectively resolve the multiple components, structure of the disks/tori and the complex evolution of an UCHII region through accreting tori. Indeed, understanding this complex phase of switching over from accretion to an HII region is crucial to a conclusive theory of the formation of massive stars. Near-infrared observations, while offering a higher spatial resolution are still prone to extinction problems and only display a partial picture. The next generation interferometers such as CARMA and ALMA can surpass these spatial resolution problems and throw much light on the details of the near environment of massive protostars and associated protoclusters. Similarly the integral field spectrometers recently operational in the near-infrared regime such as the UIST/IFU on UKIRT and SINFONI instrument on VLT can be used to clearly model the physical components of ionised gas and molecular gas by using multiple lines of molecular and atomic hydrogen as tracers. Therefore the next generation telescopes offer the opportunity to conduct detailed observational studies and modelling of the exact mechanism of massive star formation and cluster formation.

## Acknowledgments

It is a pleasure to thank Drs. Bachiller, Davis, Ojha, Shepherd, Tafalla and Mr.Eoin Clerkin who have contributed significantly to the research work described here. This article has benefited from useful discussions with Eric Keto during the workshop. I also thank Leonardo Testi for sharing data on Mol148.

## References

Adams, F. C., & Myers, P. C. (2001). *ApJ* 553:744 (AM01)

Behrend, A. & Maeder, A. (2001). *ApJS* 373:190

Beuther, H., Schilke, P., Sridharan T. K., Menten K. M., Walmsley C. M., & Wyrowski F. (2002a). *A&A* 383:892

Beuther, H., Schilke, P., Sridharan T. K., Menten K. M., Walmsley C. M., & Wyrowski F. (2002b). *A&A* 383:892

Beuther, H., Schilke, P. & Gueth, F. (2004). *ApJ* 608:330

D' Antona, F., & Mazitelli, I. (1994). *ApJ* 90:467

Elmegreen, B. G., Efremov, Y., Pudritz, R., & Zinnecker, H. (2000). in *Protostars & Planets IV eds., Mannings, V., Boss, A. P. & Russell, S. S., University of Arizona Press* 179

Herbig, G. H., (1962). *ApJ* 135:736

Hillenbrand, L., (1995). *PhD Thesis, University of Massachusetts*

Hillenbrand, L. A. & Hartmann, L. W. (1998). *ApJ* 492:540

Keto, E., (2003). *ApJ* 599:1196

Kumar, M. S. N., Bachiller, R., & Davis, C. J. (2002). *ApJ* 576:313

Kumar, M. S. N., Ojha, D. K., & Davis, C. J. (2003). *ApJ* 598:107

Kumar, M. S. N., Fernandes, A. J. L., Hunter, T. R., Davis, C. J. & Kurtz, S. (2003). *A&A* 412:175

Kumar, M. S. N., Kamath, U. & Davis, C. J. (2004). *MNRAS* 353:1025

Kurtz, S., Cesaroni, R., Churchwell, E., Hofner, P., & Walmsley, C. M. (2000). in *Protostars & Planets IV eds., Mannings, V., Boss, A. P. & Russell, S. S., University of Arizona Press* 299

Lada, C. J. & Adams, F. C. (1992). *ApJ* 393:278

Lada, C. J. & Lada, E. A. (2003). *ARA&A* 41:57 (LL03)

Lada, E. A. & Lada, C. J. (1995). *AJ* 109:1682

Li Zhi-Yun. & Nakamura, F. (2002). *ApJ* 578:256 (LN02)

Meynet, G. & Maeder, A. (2000). *A&A* 361:101

Molinari, S., Brand, J., Cesaroni, R., & Palla, F. (1996). *A&A* 308:573 (MBCP96)

Molinari, S., Brand, J., Cesaroni, R., Palla, F., & Palumbo, G. G. C. (1998). *A&A* 336:339

Molinari, S., Brand, J., Cesaroni, R., & Palla, F. (2000). *A&A* 355:617

Muench, A. A., Alves, J. F., Lada, C. J., & Lada, E. A. (2001). *ApJ* 558 (L51)

Muench, A. A., Lada, E. A., Lada, C. J., & Alves, J. F. (2002). *ApJ* 573:366

Meyer, M. R., Calvet, N., & Hillenbrand, L. (1997). *AJ* 114:288

Palla, F., Brand, J., Comoretto, G., Felli, M., & Cesaroni, R. (1991). *A&A* 246:249

Shepherd, D. S., & Churchwell, E. (1996). *ApJ* 472:225

Shepherd, D. S., Kumar, M. S. N., & Tafalla, M. (2005). in preparation

Sridharan, T. K., Beuther, H., Schilke, P., Menten, K. M., & Wyrowski, F. (2001). *ApJ* 566:931 (Sri02)

Testi, L., Palla, F., Prusti, T., Natta, A., & Maltagliti, S., (1997). *A&A* 320:159

Testi, L., Palla, F. & Natta, A. (1999). *A&A* 342:515

Yorke, H. & Sonnhalter, C. (2002). *ApJ* 569:846

Wood, D. O. S. & Churchwell, E. (1989). *ApJ* 340:265 (WC89)

# PRE-MAIN-SEQUENCE EVOLUTION AND BROWN DWARFS BEYOND THE SOLAR VICINITY*

A. Moitinho [1], C.J. Lada [2] , N. Huélamo [3], J.F. Alves [4], A.A. Muench [2]

[1] *CAAUL Observatório Astronómico de Lisboa, Tapada da Ajuda, 1349-018 Lisboa, Portugal*

[2] *Harvard-Smithsonian Center for Astrophysics, Mail Stop 72, Cambridge, MA 02138*

[3] *ESO, Santiago de Chile, Chile*

[4] *ESO, Karl Schwarzschild Straß e 2, D-85748 Garching, Germany*

andre@oal.ul.pt,clada@cfa.harvard.edu, nhuelamo@eso.org,jalves@eso.org, gmuench@cfa.harvard.edu

**Abstract** Most of what is known about pre-main-sequence (PMS) evolution and brown dwarfs (BD) is based on results for nearby (up to a few hundreds of parsecs) star formation regions. Although close, observations of these regions are in general compromised by variable amounts of dust and nebular contamination which affect not only the derived properties of individual objects, but also bias the samples of members. Avoiding these problems forces us to look out to farther regions, something that is now becoming possible with the new generation of large telescopes.

In this contribution, we present new VLT observations in the V and I bands which uncover the low mass PMS stars and BDs of the 5 Myr open cluster NGC 2362. This cluster is located at 1480 pc and is not affected by the problems mentioned above, which makes it a privileged target for studying not only PMS evolution and BDs but also offers a valuable opportunity to study the IMF well down into the substellar regime.

## Introduction

Star clusters are privileged laboratories for stellar astrophysics. Interpretation of star cluster color-magnitude diagrams (CMDs) has led to the understanding of fundamental aspects of stellar structure and evolution. For the youngest clusters, the CMDs provide insight to the evolution of their, newly formed, contracting pre-main-sequence stars. Also, since clusters represent a complete star formation event, the detailed shape and possible variation of their

*Based on observations collected at the European Southern Observatory, Paranal, Chile (ESO Programmes 70.C-0448 and 072.C-0765)

Initial Mass Function (IMF) with chemical composition, location, or time, not only provides important clues to the star formation process itself, but is also a key ingredient for understanding the dynamic and chemical evolution of all stellar systems, from clusters to galaxies.

However, care must be taken when deriving a cluster IMF. Because dynamical evolution disperses clusters and produces mass segregation, obtaining a complete census of a cluster's members can prove to be tricky. Indeed, trying to compensate for these effects by covering large areas in the sky is not only very time consuming from the observational point, but also increases the number of field interlopers, adding noise to the derived luminosity and mass functions. So avoiding the effects of dynamical evolution brings us to the youngest clusters, which will host a pre-main-sequence population. The study of the IMF in its low mass end will further benefit from using PMS stars, since they can be much brighter than their more evolved counterparts. Observationally, however, these stars are among the hardest to study because they are usually heavily extincted by their parental molecular cloud and are often seen in projection against bright HII regions.

The ideal laboratories to study PMS evolution and the IMF would then be the youngest possible galactic clusters free from dust extinction and nebula contamination, at a distance that permits detection of their low-mass mass stars and brown dwarfs. An investigation of such an ideal cluster would provide robust observables to test pre-main-sequence models (luminosity functions, color-magnitude, and color-color diagrams), free from completeness corrections and scatter due to variable extinction and nebula.

The southern young open cluster NGC 2362 ($\alpha_{2000} = 07^h18^m46^s.3$, $\delta_{2000} = -24°57'22''$) is perhaps one of the best possible examples of the ideal cluster described above. The cluster is located at 1480 pc and is virtually free from dust extinction and shows no signs of nebular emission.

Historically, NGC 2362 has been used as the standard observational template to define the upper end of the Zero Age Main Sequence (Sandage, 1956).

Recently, the multicolor surveys of NGC 2362 performed by Moitinho et al., 2001 and Moitinho et al., 2003, revealed a long and well defined pre-main-sequence, spanning about 12 magnitudes in the $V$ vs. $V - I$ color-magnitude diagram, from A-stars down to approximately $50\mathrm{M}_{\mathrm{Jup}}$, uncovering an important population of brown dwarfs. Analysis of the color-color and color-magnitude diagrams show that the cluster is virtually free of dust and nebular contamination and confirm previous distance estimates. Detailed comparisons of the cluster high-mass and pre-main-sequence populations with evolutionary models allowed to obtain the first reliable age estimate (5Myr) for this cluster.

More extended overviews of the previous knowledge on NGC 2362, and discussions of the cluster's reddening, distance and age can be found in Moitinho et al., 2001 and Moitinho et al., 2003.

In this contribution we present new VLT optical data of NGC 2362 which uncover the cluster's PMS down to approximately 30 Jupiter masses, well below the hydrogen burning limit. The Luminosity Function (LF) and a preliminary IMF of NGC 2362 reaching well into the substellar regime are also shown for the first time.

## Observations

The optical data presented in this contribution was acquired in 3 runs.

UBVRI imaging data of NGC 2362 and of a nearby control field were obtained in February 2001 using DFOSC mounted on the Danish 1.5m in La Silla. The observational difficulties imposed by the forth magnitude central star $\tau$CMa (severe bleeding, scattered light, ghost images) were by-passed by pointing at four positions (NSEW) as close as possible to $\tau$CMa but avoiding it or its scattered light (see left panel of Fig. 1). A series of short exposures (1 sec) centered on $\tau$CMa were also acquired. Complete details on these observations and reductions are given in Moitinho et al., 2001.

The two other runs were carried out at the VLT, in Paranal, in Jan 2003 (one night in visitor mode) and Jan 2004 (service mode). Dithered V and I band exposures centered on NGC 2362 and on the control field were acquired with the FORS camera. This time, data corruption produced by $\tau$CMa (and a couple of other bright stars in the field) was largely avoided by using the occulting fingers of the MOS unit. The MOS was rotated so that the brightest stars could be covered by fingers coming from the left (east) side or from the right (west) side of the frame. Half the exposures were taken in each configuration. The images were then registered and average-combined. Data underneath the occulting fingers were rejected by the averaging task (iraf/imcombine). The final $V$ image is shown in middle panel of Fig. 1. The total integration times for each field were about 5 hours in V and 6 minutes in I.

The PSF photometry of the VLT observations was tied to the photometric catalog obtained with the 1.5m and described in Moitinho et al., 2001. Measurements from both sources were then average combined using their errors as weights.

## Results

## The color-magnitude diagram

Fig. 2 introduces the $V$ vs. $V - I$ color-magnitude diagrams (CMDs) of NGC 2362 (left panel) and of the nearby control field (CF; middle panel). Only data within the VLT observations' box and with errors smaller than 0.05 mag have been plotted. The area covered by the CF data is the same as for the cluster. It is readily seen that the red sequence that extends from $V \sim 15$ to 26

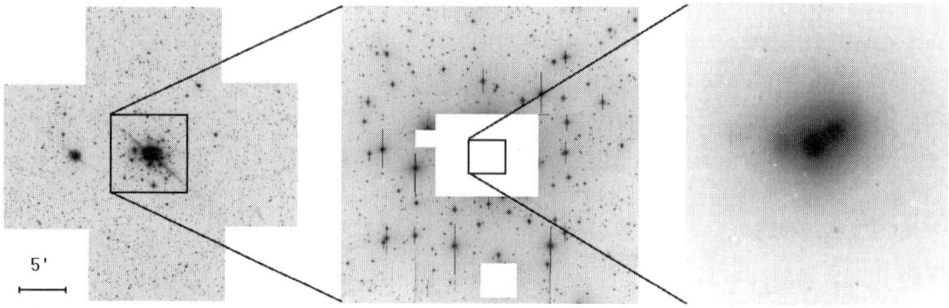


*Figure 1.* **Left**: 1.5m data - Mosaic of V band images of NGC2362. North is up and east is left. The surveyed area covers $\sim 540'^2$. The central $\sim 1.2'^2$ portion of the mosaic, which includes $\tau$CMa, is composed of a 1s exposure. **Middle**: VLT FORS image - Combined exposures in the V band. The fingers in the MOS unit were used as masks to occult the brightest stars in the field (see text for details). **Right**: VLT NACO K band image of the O9Ib supergiant multiple star $\tau$CMa.

has no counterpart in the CF's CMD. This red branch is part of a very long and well defined 5 Myr pre-main-sequence (PMS) band that starts at about $V \sim 12$ and is clearly visible from $V \sim 15$ to $V \sim 26$. Many of the sources along this PMS strip have been found by Huélamo et al., 2003 to also have X-ray emission, further assessing their PMS nature.

On the right panel of Fig. 2, a number of theoretical 5 Myr isochrones have been superimposed, shifted to account for the effects of reddening and distance ($E(B-V) = 0.1$ mag; $(v - M_v) = 11.16$ mag). A thorough discussion of the reddening, distance and age determinations are given in Moitinho et al., 2001. The dotted line at $V \sim 20.5$ indicates the limiting magnitude for the 1.5m data. The dashed lines at $V \sim 23.1$ indicate the range for the hydrogen burning limit (HBL), $M = 0.08M_\odot$, depending on the adopted theoretical model . Similarly, the fainter solid lines plotted indicate the magnitude of a 30 Jupiter masses brown dwarf.

The bright portion of the isochrones nicely fits the observed sequence and has allowed to constrain the age of NGC 2362 as discussed in Moitinho et al., 2001. On the other hand, the low mass/cool stars are not well fitted by any of the 5 Myr isochrones. This results from the well known fact that the model predictions of optical colors are not robust at cool temperatures (Baraffe et al., 1998). In particular, the often cited problems with molecular line lists as TiO seem indeed to be responsible for part of the deviations, as evidenced by the not-so-bad agreement of the models using improved line lists (the IMPR model marked on Fig. 2). However, the much better agreement of the COND models suggests that the problem comes from much more basic ingredients in the

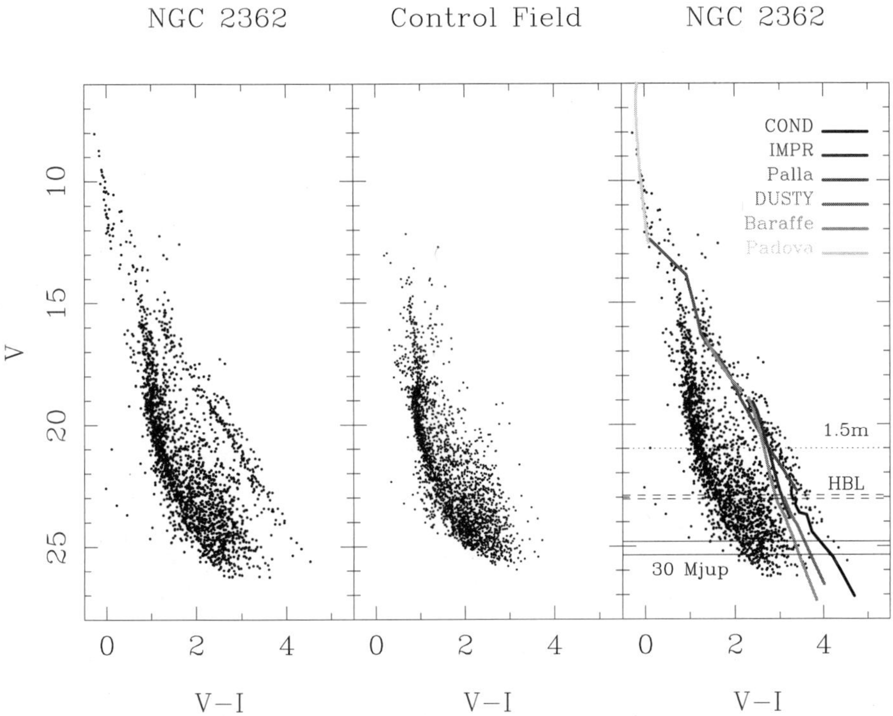


*Figure 2.* **Left**: Deep color-magnitude diagram (CMD) of NGC 2362. **Middle** Deep CMD of the control field: **Right**: The same as in the left panel but with several theoretical 5Myr models superimposed. The isochrones were calculated using a solar recipe and have been shifted to account for the effects of reddening and distance. The models are: Post main sequence from Girardi et al., 2000 (Padova); PMS from Baraffe et al., 1998 (Baraffe); PMS with improved TiO line lists from Baraffe (priv. comm. - IMPR); Chabrier et al., 2000 dusty (DUSTY) and Baraffe et al., 2003 settled dust (COND) PMS models; 3Myr PMS model of Palla and Stahler, 1999 .

models, namely the way dust is treated. Since such effects aren't severe at infrared wavelengths - the range of election for most star formation studies - this is another point where the present optical data, uncontaminated by the environment, can be a rare and extremely useful dataset.

Although not as consistent with dynamical masses as the models of Baraffe et al., 1998, the isochrone of Palla and Stahler, 1999 apparently reproduces the observed sequence much better. However, the transformation from the theoretical to the observational plane is empirical, which greatly hides the model's details, so it is natural that the colors will be a better match than the theoretical ones of Baraffe et al., 1998.

The cluster sequence in the CMD delineates a PMS that reaches down below the HBL, to about 30 Jupiter masses. The remarkable separation between the cluster and the field allows us to easily identify $\sim$15 BD candidates. Even if there are some field interlopers, the BD sequence is so obvious and so well separated from the field that most of the candidates are likely to be real BDs. Fig. 1 shows the spatial distribution of BDs overplotted on the VLT I band combined image. The BDs are spread out across the whole frame, showing no particular concentration. Because these active optics images are very sharp, the BD candidates are detected even close to heavily saturated stars, so the main source of BD incompleteness is the signal-to-noise detection limit and is little affected by source confusion (evidently, we are not considering the masked region).

That the PMS is so tight is indicative of a very simple star formation history, most probably a single and quick episode of star formation. Because this cluster is affected by little and non-variable extinction, as indicated by the tight PMS in the color-magnitude diagram, no complicated corrections (effects of high and/or variable extinction) had to be applied to the observed sequence. These data, therefore, provide a template that evolutionary models must match.

## The luminosity and mass functions

As mentioned in the beginning of this contribution, the spatial distribution of cluster members is of direct relevance for determining the cluster Luminosity and Initial Mass Functions. Fig. 4, shows the spatial number density of stars roughly along the cluster sequence in the CMD. Only stars within the completeness limits of the 1.5m data($V < 20$ or $0.35M_\odot$) were used in order to avoid artifacts introduced by the deeper, but spatially more limited, VLT data. The density was calculated by convolving the surface stellar distributions with a 100x100 pix (40"$\times$40") gaussian kernel, in 50 pix steps, as described in Moitinho et al., 1997. Density contours were drawn at 10, 20 and 30 stars/arcmin$^2$.

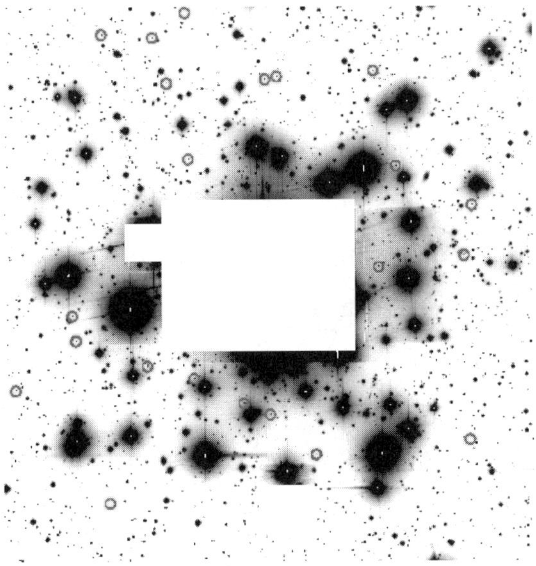

*Figure 3.* VLT combined I band image showing the spatial distribution of brown dwarf candidates. The BDs are marked with circles.

The X and Y cuts presented in 4, show that selection of stars along the cluster sequence has left only a residual number of fore/background stars. Most of these stars have magnitudes in the range $13 < V < 16$ where the cluster and field populations are not so well separated in the CMD. The dashed ellipse marks the derived cluster limits. It is centered on $\alpha_{2000} = 07^h18^m40^s.9$, $\delta_{2000} = -24°56'54''$ and has 8.9'x 11.1' axes. The solid square marks the area covered by the VLT observations. The figure shows that the VLT data cover most of the cluster area, so corrections for dynamical effects on the derived LF and IMF should be small.

The V band LF presented in the left panel of 5 was built in 0.5 mag bins using the stars close to the cluster sequence in the CMD. Contamination from field stars was compensated by subtracting the distribution of control field stars located within the same limits in the CMD as the selected cluster field stars. Care was taken to use the same area for cluster and control field data. In this figure, the dashed line marks the HBL and the dotted line marks the completeness limit. The main features of this LF are the prominent Wielen gap which is evident around $M_V \sim 3$ mag, and the peak at $M_V \sim 8.75$. The LF also shows something that was already clear in the CMD, that the ratio of BD-to-stars in the cluster is around 5%. Even considering that completeness is cutting out BDs, the shape of the low-mass end of the LF strongly indicates that the number of missing BDs should be very small.

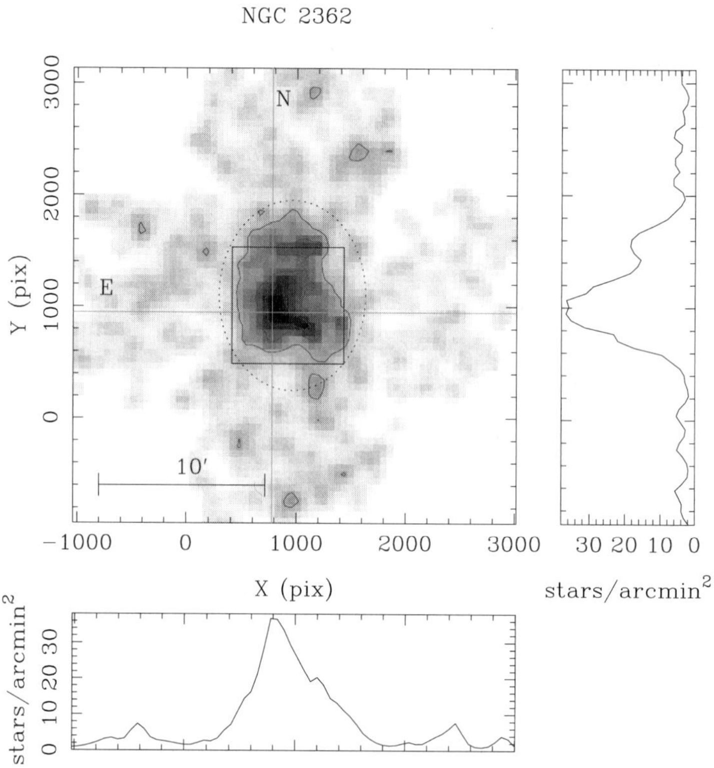


*Figure 4.* Stellar density map of NGC 2362 derived from the 1.5m data. The contours are drawn at 10, 20 and 30 stars/arcmin$^2$. The dashed ellipse marks the adopted cluster limits. The solid square marks the area covered by the VLT observations.

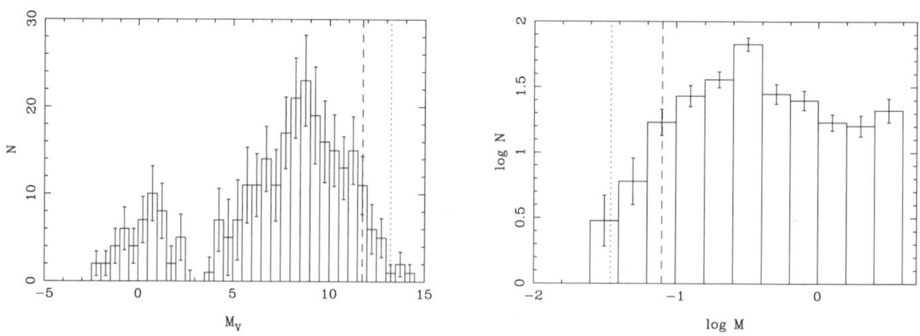


*Figure 5.* **Left**: V Luminosity function of NGC 2362. **Right**: IMF derived with mass-luminosity relations from Girardi et al. (ZAMS), Palla & Stahler (for PMS stars with $M > 1.2M_{\odot}$) and Baraffe et al. "standard" models ($M < 1.2M_{\odot}$). In both panels, the dashed line marks the HBL and the dotted line marks the completeness limit.

Possibly, the best way to derive the cluster IMF is to simulate its LF using montecarlo techniques as in Muench et al., 2000. Although this more sophisticated determination of the IMF is still underway, a quick look IMF is easily obtained by building the distribution of logarithmic masses (in solar units).

One of the key ingredients in building the IMF is the mass-luminosity relation. For the IMF presented in the right panel of Fig. 5, we employed the relations from the models of Girardi et al., 2000 (for the ZAMS stars), Palla and Stahler, 1999 (for PMS stars with $M > 1.2M_{\odot}$) and Baraffe et al., 1998 "standard" models (for PMS stars with $M < 1.2M_{\odot}$).

In the range where the data is complete, the derived IMF spans from early B stars to $\sim$50$M_{\rm Jup}$ BDs. An evident peak appears at 1/3 of $M_{\odot}$, after which the IMF falls. Other combinations of the models shown in Fig. 2 were tested (also using the I band LF) and yielded IMFs with very similar shapes but with measurable differences for the peak mass, ranging from 1/2 to 1/3 of a solar mass. So, even being preliminary, the derived IMF leaves no doubt that it falls toward the lowest masses.

## Summary and final remarks

The 5 Myr open cluster NGC 2362 is virtually free from dust and nebular contamination. Located at a distance of 1480 pc, it is close enough so that the largest modern telescopes can observe its faint substellar population. On the other hand, it is not so close as to cover a large area in the sky, increasing the contamination by field stars and making observations with small CCD fields unpractical.

The new VLT optical data presented in this work have revealed a very long and well defined PMS covering about 14 magnitudes in the $V$ vs. $(V - I)$ CMD, from early A stars to about 30 Jupiter masses. The good separation between the PMS and the field has allowed a straightforward identification of members down into the substellar regime revealing 15 likely brown dwarfs.

That the PMS is so well defined is a result of non-variable reddening and a single quick star forming episode. These unique characteristics of the dataset provide strong constraints to contemporary models of young low mass stars and brown dwarf evolution. In particular, these highly uncontaminated optical data can be extremely valuable for constraining the way dust is treated in the models.

Analysis of spatial structure of stars selected roughly along the sequence defined by NGC 2362 in the CMD has allowed to derive the clusters' limits (an ellipse with 8.9'x 11.1' axes) and to verify that this rough selection provided a sample largely free from field star contamination.

The V band LF and a preliminary IMF were built. Inspection of the LF showed that the ratio of BD-to-stars in the cluster is around 5%. The derived

IMF spans from early B stars to $\sim 50 M_{Jup}$ BDs, with an evident peak at 1/3 of $M_{\odot}$ after which the IMF falls.

Using also different sets of models and I band data, it was found that the derived IMFs had very similar shapes but that the peak mass varied from 1/2 to 1/3 of a solar mass. The present dataset leaves no doubt that the IMF falls toward the lowest masses around 0.3-0.5 solar masses.

Overall, NGC 2362 may provide the most homogeneous and complete set of PMS stars for an individual star formation region yet.

## Acknowledgments

This research has used the ADS, WEBDA and DSS databases. A.M. acknowledges financial support by FCT (Portugal; grant BPD/20193/99).

## References

Baraffe, I., Chabrier, G., Allard, F., and Hauschildt, P. H. (1998). Evolutionary models for solar metallicity low-mass stars: mass-magnitude relationships and color-magnitude diagrams. *A&A*, 337:403.

Baraffe, I., Chabrier, G., Barman, T. S., Allard, F., and Hauschildt, P. H. (2003). Evolutionary models for cool brown dwarfs and extrasolar giant planets. The case of HD 209458. *A&A*, 402:701.

Chabrier, G., Baraffe, I., Allard, F., and Hauschildt, P. H. (2000). Evolutionary Models for Very Low-Mass Stars and Brown Dwarfs with Dusty Atmospheres. *ApJ*, 542:464.

Girardi, L., Bressan, A., Bertelli, G., and Chiosi, C. (2000). Evolutionary tracks and isochrones for low- and intermediate-mass stars: From 0.15 to 7 $M_{\odot}$, and from Z=0.0004 to 0.03. *A&AS*, 141:371.

Huélamo, N., Stelzer, B., Moitinho, A., Alves, J. F., and Lada, C. J. (2003). X-Ray emission from NGC 2362. In Lépine, J. R. D. and Gregorio-Hetem, J., editors, *Open Issues in Local Star Formation*, volume 299 of *A&SSL*, page 194. Kluwer.

Moitinho, A., Alfaro, E. J., Yun, J. L., and Phelps, R. L. (1997). CCD UBV Photometry of the Young Open Cluster NGC 3766. *AJ*, 113:1359.

Moitinho, A., Alves, J., Huélamo, N., and Lada, C. J. (2001). NGC 2362: A Template for Early Stellar Evolution. *ApJL*, 563:L73.

Moitinho, A., Lada, C. J., Huélamo, N., and Alves, J. F. (2003). X-Ray emission from NGC 2362. In Lépine, J. R. D. and Gregorio-Hetem, J., editors, *Open Issues in Local Star Formation*, volume 299 of *A&SSL*, page 47. Kluwer.

Muench, A. A., Lada, C. J., and Lada, E. (2000). Modeling the Near-Infrared Luminosity Functions of Young Stellar Clusters. *ApJ*, 553:338.

Palla, F. and Stahler, S. W. (1999). Star Formation in the Orion Nebula Cluster. *ApJ*, 525:772.

Sandage, A. (1956). The Systematics of Color-Magnitude Diagrams and Stellar Evolution. *PASP*, 68:498.

# BROWN DWARFS

R. Rebolo
*Instituto de Astrofísica de Canarias*
*38200, La Laguna, Tenerife, Spain*
rrl@iac.es

**Abstract** Hundreds of brown dwarfs have been discovered since 1995. These objects are likely to populate every region of our Galaxy and possibly other galaxies in comparable number to stars. We briefly review their basic properties, mass function, spatial distribution and possible formation scenarios.

## Introduction

Brown dwarfs span the mass range between massive giant planets and the lowest mass stars. The frontier between stars and brown dwarfs is set by the minimum mass for stable hydrogen burning in stellar interiors, which for solar metallicity corresponds to $\sim$ 73 $M_{Jup}$ (Baraffe et al. 1998). The frontier with the planetary domain is much more subtle. Ideally, the distinction between the lowest mass brown dwarfs and the more massive giant planets discovered in radial velocity surveys could be established on the basis of a different formation mechanism. If brown dwarfs were formed only as a result of the fragmentation of molecular clouds, i.e. like stars, and massive giant planets were only formed in protoplanetary discs via gravitational instability, accretion of planetesimals or any other mechanism, we could possibly expect different physical conditions in the interiors of these two types of objects and establish a distinction between them. However, the poor knowledge of the formation history of such low-mass objects, and the lack of information on the physical conditions of their interiors prevent a useful criterium to distinguish them. For the sake of clarity, we will adopt as an arbitrary limit between brown dwarfs and planetary-mass objects the minimum mass for deuterium burning: $\sim$ 12 $M_{Jup}$, for solar metallicity.

On a theoretical basis, the radius of Jupiter is slightly larger than the radius of an old brown dwarf and consequently, the density of brown dwarfs would be 15-80 times larger than the density of Jupiter. Brown dwarf luminosities are very sensitivy to evolutionary stage, progressively decreasing with age. Old brown dwarfs (age > 1 Gyr) are expected to be $10^4$- $10^6$ times less luminous than the Sun and display atmospheric effective temperatures below 1000 K. The lowest mass brown dwarfs may reach effective temperatures as low as 200 K at sufficiently old ages.

The existence of brown dwarfs was speculated by Kumar (1963) already in the 60s, but it took more than 30 years to find convincing examples of these elusive objects. In spite of active searches in young star clusters where they were expected to be still contracting and hence to be much more luminous, no brown dwarf had been recognized before 1995. Searches in the Pleiades (Jameson and Skillen 1989, Stauffer et al. 1989,1994), in $\alpha$ Persei (Rebolo et al. 1992), $\rho$ Ophiuchi (Rieke and Rieke 1990) or in Taurus (Stauffer et al. 1991) produced a number of candidates. However, the effective temperatures and luminosities of these candidates were too high to ensure a substellar nature. One of the intriguing brown dwarf candidates, GD 165B, was actually discovered around a white dwarf (Zuckermann and Becklin 1992). Its peculiar spectral energy distribution later recognized as characteristic of L-type dwarfs, could not be taken as a definitive proof of substellar nature since sufficiently low-mass stars end with these spectral types. The Li test (Rebolo et al. 1992) showed that many of the coolest field dwarfs considered at that time good brown dwarf candidates, had destroyed a large fraction of their initial Li content, and therefore were most likely stars (Magazzù et al. 1993, Martín et al. 1994).

In 1995, unambiguous brown dwarfs were discovered in the Pleiades star cluster and orbiting around the star Gl 229. The search for fainter objects in the Pleiades led to the detection of a cool and faint proper motion member of the cluster: Teide 1 (Rebolo et al. 1995) whose M8 spectral type and low luminosity could only be explained if the mass was strictly below the substellar mass limit. This detection was closely followed in time by the finding of a twin object (Calar 3) also in the Pleiades and the confirmation that both had Li in the atmosphere (Rebolo et al. 1996). The positive result of the Li test, fully supported the claimed brown dwarf nature of Teide 1. Adaptive optics searches around stars in the solar neighbourhood led to the discovery of a very cool and low luminosity brown dwarf orbiting the star Gl 229 B (Nakajima et al. 1995). These first examples of young and old brown dwarfs with masses in the range 40-50 $M_{Jup}$ have been followed in subsequent years by the discovery of hundreds of brown dwarfs in star clusters and in the field (in much lower numbers also around stars). There is evidence that brown dwarfs form with any mass between the hydrogen and the deuterium burning limits. There are examples of brown dwarfs with effective temperatures down to 700 K and the standard spectral classification has been expanded in two new spectral types in order to account for the spectroscopic richness of these objects. An increasing number of substellar mass function determinations mostly in stellar clusters show that brown dwarfs are extremely common in the Galaxy and hint they may outnumber stars. We review here, the status of the searches in star clusters and associations, in the field and around stars. We also briefly discuss recent results on the multiplicity of these objects and various formation mechanisms proposed to explain the main observational features.

## Brown dwarfs and new Spectral types

The first field brown dwarfs (Ruiz et al. 1997, Delfosse et al. 1997) and subsequent detections of brown dwarfs around stars (Rebolo et al. 1998) showed objects with spectral characteristics between those of young brown dwarfs in clusters and the very cool Gl 229 B. These findings and the discovery of tens of such ultracool objects by the 2MASS survey (Kirkpatrick et al. 1999, Reid et al. 1999) led to establish the new spectral class L (Martín et al. 1997, 1999; Kirkpatrick et al. 1999). L-dwarfs are characterised by the disappearance of the red TiO and VO bands (typical of M-dwarfs) from the optical spectrum, the increasing dominance of broad absorption resonance lines of Na I and K I, strong $H_2O$ absorption bands and persistent CO overtone bands at 1-2.5 $\mu$m. Their near infrared colours follow $J - K > 1.3$ and $I - J > 3$, and their effective temperatures can be estimated from the following relationship with spectral type $T_{eff}$ = (2380±40) - (138±8) SpT where SpT ranges from 0 to 8 for spectral types L0 to L9 (Burgasser 2001).

For about four years Gl 229 B remained in a class by itself, no other object with similar spectral characteristics was found till 1999 when various searches proved the existence of such analogues and showed their space densities were similar to L dwarfs (Burgasser et al. 1999, 2000; Cuby et al. 1999, Leggett et al. 2000). A new T spectral classification (see e.g. Burgasser et al. 2002) was proposed for these objects which are characterised by strong $CH_4$ absorptions in the $H$ and $K$ bands, strengthening of water bands as compared with L dwarfs, and blue infrared colours $J - K \sim 0$ with large increase in flux from the red part of the optical spectrum to the near infrared ($I - J > 4$). A relationship between effective temperature and spectral type derived from the K-band bolometric correction shows a monotonic behaviour throughout the L-T sequence (Nakajima et al. 2004). T dwarfs are classified from T0 to T8. The effective temperatures of the latest T dwarfs are in the range 900-800 K. To date about 40 T dwarfs have been discovered, but the edge of the *stellar* mass function in the field, or in stellar clusters, has not been reached yet and it is certainly expected that even cooler objects exist down to the temperatures of the Jovian-like planets. In the following sections, we review our knowledge about the spatial and mass distribution of brown dwarfs around stars, in the field and in stellar clusters.

## Companions to stars

Radial velocity surveys have shown the existence of more than 130 planets around solar type stars and a small number of companion brown dwarfs (m > 13 $M_{Jup}$). Above 7% of solar type stars appear to have planets at separations less than 5 AU with masses M sin (i)=0.25-13 $M_{Jup}$. The mass distribution of these planets show a decreasing trend towards large masses, such that brown dwarfs are detected around less than 0.5 % of solar type stars at separations smaller than 4 AU. For comparison, stellar companions in this separation range are found around $\sim$ 10 % of the stars. The paucity of brown dwarfs at such distances, has been designated as the "brown dwarf desert". Possibly indicating a different formation mechanism for substellar objects. There are, however, some stars with *oasis* in the "desert", for example, this is the case of the star HD 137510 with a companion of Msin (i)= 26 $M_{Jup}$ orbiting at 1.85 AU (Endl et al. 2004).

At much larger separations brown dwarfs appear to be more frequent. Since the first detections of brown dwarfs around stars (Gl 229 B orbiting at $\sim$45 AU, Nakajima et al. 1995; and G 196-3 B at $\sim$ 300 AU, Rebolo et al. 1998), more than a dozen other brown dwarfs have been directly imaged as wide companions. Remarkably, some of them are found at very large distances of their stars. For example, Scholz et al. (2003) reported the discovery of $\epsilon$ Indi B, a $\sim$ 50 $M_{Jup}$ T2.5 brown dwarf as a very wide companion ($\sim$ 1500 AU) of the nearby high proper motion star $\epsilon$ Indi A, at a distance of 3.626 pc from the Sun. About 1% of the stars seem to have brown dwarf companions at separations larger than 75 AU (McCarthy and Zuckerman 2004), and about 1 % of early M dwarfs appear to have them at separations between 10 and 50 AU (see e.g., Forveille et al. 2004). The discovery of a brown dwarf orbiting the M8 dwarf LHS 2397a (Freed et al. 2003, Masciadri et al. 2003) at separation of 2.34$\pm$ 0.14 AU suggests that the frequency of brown dwarfs around very low mass stars is higher. For comparison, the frequency of M dwarf stellar systems is about 35 % with a peak in the range 3-30 AU. Efforts to uncover the mass and radial distribution of extra-solar planets around M stars have also started to reveal super Jovian-mass planets within a few AU of their central stars. These searches are at a too early stage to establish any significant comparison with the frequency of brown dwarfs in similar stars.

## Searches in star clusters and young stellar associations: the brown dwarf mass function

The higher luminosity of brown dwarfs at early evolutionary stages have allowed to explore a low mass domain in young star clusters which is still elusive in large scale field searches, simply because the average brown dwarf population is significantly older and intrinsically fainter. The first substellar studies in the Pleiades cluster showed a well populated brown dwarf sequence in colour-magnitude diagrams down to $\sim$ 35 $M_{Jup}$ (Zapatero-Osorio et al. 1997, Bouvier et al. 1998, Martín et al. 1998). The determination of the mass function N(M)$\propto$ $M^{-\alpha}$ by these authors led to values in the range 0.4-1 for the spectral index. Recent surveys (Nagashima et al. 2003, Jameson et al. 2002, Moureaux et al. 2003) extend further and confirm these findings. A more recent wide near-IR search for brown dwarfs in the cluster covering $\sim$ 1.5 square degrees (Bihain et al. 2005, in preparation) has found proper motion confirmed Pleiades brown dwarfs with I-J and J-K colours typical of late L dwarfs. The faintest objects found may have masses close to 25 $M_{Jup}$ if the age of the cluster is 120 My.

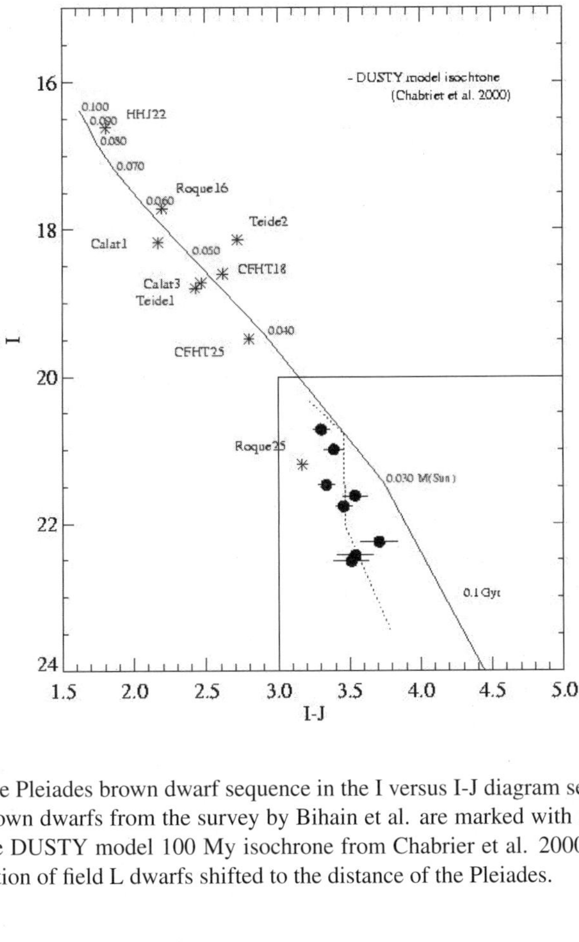


*Figure 1.* The Pleiades brown dwarf sequence in the I versus I-J diagram sequence. The new candidate L brown dwarfs from the survey by Bihain et al. are marked with filled circles. The solid line is the DUSTY model 100 My isochrone from Chabrier et al. 2000. The dotted line shows the location of field L dwarfs shifted to the distance of the Pleiades.

Searches for even less massive brown dwarfs have been conducted in many other younger clusters and stellar associations. These searches have shown the ubiquituous presence of brown dwarfs with rather large numbers. The mass spectrum has been determined down to 20-30 $M_{Jup}$ in clusters with a variety of ages and stellar densities: for example, in $\rho$ Ophiuchi (Williams et al. 1995, Luhman et al. 2000), Chamaeleon I (Comerón et al. 2000), Trapezium and IC 348 (Luhman et al. 2000) and $\sigma$ Orionis (Béjar et al. 1999, 2001). The substellar initial mass functions are roughly similar in spite of a range of more than 2 orders of magnitude in the density of star formation. Interestingly, the faintest substellar objects discovered in these surveys have masses below 10 $M_{Jup}$. Zapatero-Osorio et al. (2000) reported on very faint, extremely red objects in the $\sigma$ Orioinis cluster with L-type spectra, the faintest would have masses close to 5 $M_{Jup}$. Lucas and Roche (2000) proposed photometric candidates to planetary-mass objects in the Trapezium cluster. Zapatero-Osorio et al. (2002) claimed the detection of a $3^{+5}_{-2} M_{Jup}$ free-floating planet in the $\sigma$ Orionis cluster with T spectral type. This is the first methane object ever discovered in a stellar cluster. It has been argued by Burgasser et al. (2004) that it may be a foreground T dwarf not related to the cluster, but the extreme I-K colour is not found in field T dwarfs which favours the cluster membership hypothesis. More recently, Caballero et al. (in preparation) have extended the survey in $\sigma$ Orionis detecting candidates of J=20-20.8 with colours typical of L6-L8 dwarfs. This suggests that T dwarfs will show J$\sim$21-21.5 and that S Ori 70 would more likely be a binary system. Béjar et al. (2001) find an index $\alpha$=0.8$\pm$0.25 for the substellar mass function in $\sigma$ Orionis down to $\sim$5 $M_{Jup}$. At lower masses, the incompleteness of the surveys prevent to reach any firm conclusion, the current data are however consistent with a smooth extrapolation of this mass function to even lower masses and hints that the number of isolated planetary-mass objects in the cluster may be comparable to the number of brown dwarfs.

## Multiplicity

The claim of a spectroscopic brown dwarf binary in the Pleiades (Basri and Martín 1999) has been followed by the detection of a large number (higher than 30) of binary brown dwarfs in the field (see e.g., Close et al. 2003). Binaries have been found among the L and T brown dwarfs down to masses $\sim$ 30 $M_{Jup}$. There are binaries with separations from a few to hundreds of AUs. The nearest brown dwarf binary, $\epsilon$ Ind Ba and Bb (McCaughrean 2004 et al.), is formed by two brown dwarfs of spectral types T1-6 with masses close to 50 and 30 $M_{Jup}$, respectively. The physical separation is $\sim$ 2.6 AU and orbital period $\sim$ 15 yr. There are also examples of wide binaries although they are rare. It is the case of 2MASS J11011926-7732383AB found by Luhman (2004) towards the Chamaeleon I cloud. The angular separation of these two M7/8 brown dwarfs is 1.44 arcsec which corresponds to $\sim$ 240 AU at the distance of this star forming region. About 15 % of field brown dwarfs appear to be binaries with separations in the range 5-20 AU (see e.g. Bouy et al. 2003). It has been claimed that the overall binary fraction in young clusters like the Pleiades could be as high as 50% (Pinfield et al., 2003), but this claim is based on the photometric properties of Pleiades brown dwarf candidates which are still rather uncertain. Direct imaging searches in the cluster give a binary fraction of 15% (Martín et al. 2003). The corresponding stellar binary fractions is in the range 50-60 % in solar type stars and decreases to $\sim$ 35 % in M types. Different surveys suggest that the distribution of orbital semi-major axes of brown dwarfs peaks at much smaller values than in stars.

Most remarkably, the study of substellar binaries has led to the first dynamical determination of the mass of a brown dwarf. Using Adaptive Optics techniques at the Keck telescope, Zapatero Osorio et al. (2004) have measured the orbital motion of the two components of the binary Gl 569B. The total mass of the system is 0.125$\pm$0.005 $M_{Jup}$ and the orbital period is 876.0$\pm$9.2 days. Radial velocity measurements of each of the components has allowed a determination of the masses resulting $m_1$=0.068$\pm$0.011 $M_\odot$ and $m_2$=0.057$\pm$0.011 $M_\odot$. The lightest component is clearly placed in the substellar domain; additional radial velocity measurements have the potential to reduce significantly the error bar in the mass determination, and clarify whether the more massive component is also a brown dwarf.

Deep searches around young brown dwarfs have also led to the discovery of a giant planet candidate imaged with NACO at VLT around the young M8 brown dwarf 2MASSWJ1207334-393254 (Chauvin et al. 2004). The primary is a proper motion member of the TW Hydrae Association (age 8$\pm$ 4 My) with a mass of $\sim$ 25 $M_{Jup}$. The faint candidate companion would have a mass of 5$\pm$2 $M_{Jup}$. A modest quality near infrared spectrum suggests an L spectral type. If confirmed, this object would have similar characteristics (mass and effective temperature) than the isolated planetary-mass objects discovered by Zapatero-Osorio et al. (2000) in the $\sigma$ Orionis cluster. It is important to study the frequency and distribution of objects with these masses both free-floating and around stars/brown dwarfs in order to clarify their origin.

## Formation mechanisms

Brown dwarfs may be the result of the collapse of low-mass cores ($m < 0.1\ M_\odot$) smaller and denser than those currently observed in star forming regions, or could result from the fragmentation of higher mass cores leading to unstable multiple systems which would be prevented to accrete enough mass by some mechanism. Reipurth and Clarke (2001) proposed that ejection from the gas reservoirs where star formation takes place could lead to the formation of objects unable to accrete enough mass to become stars. Photoerosion and ionization by very massive stars (Whitworth and Zinnecker 2004) may also inhibit the accretion process. Observations in multiple star forming regions with different environmental conditions (different star densities, existence of O-type stars, etc.) are required to distinguish among these scenarios. On the other hand, gravitational instability in protostellar discs could also produce low-mass brown dwarfs (see e.g. Jiang et al. 2004).

Numerical simulations of star formation to resolve the fragmentation process down to the opacity limit in stellar clusters (Bate et al. 2003) shows that star formation is a higly-dynamic and chaotic process. It occurs in localized bursts within the cloud via the fragmentation of dense molecular cloud cores and of massive circumstellar discs. Star-disc encounters form binaries and truncate discs, and stellar encounters disrupt bound multiple systems. These simulations find that brown dwarfs and stars are produced in similar numbers. Bate et al. (2003) predict a large fraction of brown dwarfs are formed via disc fragmentation as compared with fragmentation in discs in rather short scaless. Few brown dwarfs are predicted to be binary and to have circumsubstellar discs.

Observations in very young star forming regions show that the process of brown dwarf formation is rather fast. Brown dwarfs and planetary mass candidates are found in regions with less than 1 My in abundant number. There is, however, very little information on the spatial distribution and proper motions of these objects to shed light on their origin. The determination of multiplicity properties, of evolution effects in the clustering properties of brown dwarfs in star forming regions, and the characteristics of the substellar mass functions in various environments will be crucial to disentangle the formation mechanism of substellar objects.

## Acknowledgments

I thank J.A Caballero, V.J.S. Béjar and G. Bihain for valuable discussions on this topic and help with the figure.

## References

Baraffe, I., Chabrier, G., Allard, F., and Hauschildt, P. H. (1998). Evolutionary models for solar metallicity low-mass stars: mass-magnitude relationships and color-magnitude diagrams. *A&A*, 337:403.

Baraffe, I., Chabrier, G., Barman, T. S., Allard, F., and Hauschildt, P. H. (2003). Evolutionary models for cool brown dwarfs and extrasolar giant planets. The case of HD 209458. *A&A*, 402:701.

Basri, G., and Martín, E. L. (1999). PPL 15: The First Brown Dwarf Spectroscopic Binary. *AJ*, 118:2460.

Bate, M. R., Bonnell, I. A., and Bromm, V. (2000). The Formation of a Star Cluster: Predicting the Properties of Stars and Brown Dwarfs. *MNRAS*, 339:577.

Béjar, V. J. S., Zapatero Osorio, M. R., and Rebolo, R. (1999). A Search for Very Low Mass Stars and Brown Dwarfs in the Young $\sigma$ Orionis Cluster. *ApJ*, 521:671.

Béjar, V. J. S., Martín, E. L., Rebolo, R., Barrado y Navascués, D., Bailer-Jones, C. A. L., Mundt,R., Baraffe, I., Chabrier, C., and Allard, F. (2001). The substellar Mass Function in $\sigma$ Orionis. *ApJ*, 556:830.

Bouvier, J., Stauffer, J. R., Martín, E. L., Barrado y Navascués, D.,Wallace, B. and Béjar, V. J. S. (1998). Brown Dwarfs and Very Low-Mass Stars in the Pleiades Cluster: a Deep Wide-Field Imaging Survey. *A&A*, 336:490.

Bouy, H., Brandner, W., Martín, E. L. et al. (2003). Multiplicity of Nearby Free-floating Late M and L Dwarfs: HST-WFPc2 Observations of Candidates and Bona Fide Binary Brown Dwarfs. *AJ*, 126:1526.

Burgasser, A.,(2001). Ph.D. Thesis Caltech .

Burgasser, A. J., Kirkpatrick, J. D., Brown, M. E. et al. (1999). Discovery of Four Field Methane (T-Type) Dwarfs with the Two Micron All-Sky Survey. *ApJ*, 522:65.

Burgasser, A. J., Kirkpatrick, J. D., Cutri, R. M. et al. (2000). Discovery of a Brown Dwarf Companion to Gliese 570ABC: A 2MASS T Dwarf Significantly Cooler than Gliese 229 B. *ApJ*, 531:L57.

Burgasser, A. J., Kirkpatrick, J. D., McGobern, M. R, McLean, I.S., Prato, L., Reid, I.N. (2004). S Orionis 70: Just a Foreground Field Brown Dwarf?. *ApJ*, 604:827.

Chabrier, G., Baraffe, I., Allard, F., and Hauschildt, P. H. (2000). Evolutionary Models for Very Low-Mass Stars and Brown Dwarfs with Dusty Atmospheres. *ApJ*, 542:464.

Chauvin, G., Lagrange, A.-M., Dumas, C. et al. (2005). A Giant Planet Candidate near a Young Brown Dwarf: Direct VLT/NACO Observations using IR Wavefront Sensing. *A&A*, 425:L29.

Close, L. M., Siegler, N., Freed, M. and Biller, B. (2003). Detection of Nine M8.0-L0.5 Binaries: The Very Low-Mass Binary Population and its Implications for Brown Dwarf and Very Low-Mass Star Formation. *ApJ*, 587:407.

Comerón, F., Neuhäuser, R., and Kaas, A. A. (2000). Probing the brown dwarf population of the Chameleon I star forming region. *A&A*, 359:269.

Cuby, J. G., Saracco, P., Moorwood, A. F., D'Odorico, S., Lidman, C., Comerón, F. and Spyromilio J. (1999). Discovery of a faint field Methane Brown Dwarf from ESO NTT and VLT Observations. *A&A* 349:L41.

Delfosse, X., Tinney, C. G., Forveille, T. et al. (1997). Field Brown Dwarfs Found by DENIS. *A&A*, 327:L25.

Endl, M., Hatzes, A. P., Cochran, W.D. et al. (2004). HD 137510: An Oasis in the Brown Dwarf Desert. *ApJ*, 611:L1121.

Forveille, T., Ségransan, D., Delorme, P.. et al. (2004). An L0 Dwarf Companion in the brown dwarf desert, at 30 AU. *A&A* 427:L1.

Freed, M., Close, L. M. and Siegler, N. (2003). Discovery of a Tight Brown Dwarf Companion to the Low-Mass Star LHS 2397a. *ApJ*, 584:453.

Jameson, R. F. and Skillen, I. (1989). A Search for Low-Mass Stars and Brown Dwarfs in the Pleiades *MNRAS*, 239:247.

Jameson, R. F., Dobbie, P. D., Hodgkin,S. T. and Pinfield, D. J. (2002). Brown Dwarfs in the Pleiades: Spatial Distribution and Mass Function *MNRAS*, 335:853.

Jiang, I.-G., Laughlin, G. and Lin,D. N. C. (2004). On the Formation of Brown Dwarfs *RMxAA*, 21:227.

Kirkpatrick, J. D., Reid, I. N., Liebert, J. et al. (1999). Dwarfs Cooler than M: The Definition of Spectral Type "L" Using Discoveries from the 2 Micron All-Sky Survey (2MASS). *ApJ*, 519:802.

Kumar, S. S. (1963). The Structure of Stars of Very Low Mass *ApJ*, 137:1127

Leggett, S. K., Geballe, T. R., Fan, X. et al. (2000). The Missing Link: Early Methane ("T") Dwarfs in the Sloan Digital Sky Survey. *ApJ*, 519:802.

Lucas, P. W. and Roche, P. F. (2000). A population of very young brown dwarfs and free-floating planets in Orion . *MNRAS*, 314:858.

Luhman, K. L.(2004). The First Discovery of a Wide Binary Brown Dwarf. *ApJ* 614:L398.

Luhman, K. L., Rieke, G. H., Young, E. T. (2000). The Initial Mass Function of Low-Mass Stars and Brown Dwarfs in Young Clusters. *ApJ* 540:L1016.

Magazzù, A., Martín, E. L., and Rebolo, R. (1993). A spectroscopic test for substellar objects. *ApJ*, 404:L17.

Martín, E. L., Rebolo, R., and Magazzù, A. (1994). Constraints to the Masses of Brown Dwarf Candidates from the Lithium Test . *ApJ*, 436:262.

Martín, E. L., Basri, G., Delfosse, X. and Forveille, T. (1997). Keck HIRES Spectra of the Brown Dwarf DENIS-P J1228.2-1547. *A&A*, 327:L29.

Martín, E. L., Basri, G., Zapatero-Osorio, M. R., Rebolo, R. and García López, R. J. (1998). The First L-Type Brown Dwarf in the Pleiades. *ApJ*, 507:L41.

Martín, E. L., Delfosse, X., Basri, G., Goldman, B., Forveille, T., and Zapatero Osorio, M. R. (1999). Spectroscopic Classification of Late-M and L Field Dwarfs. *AJ*, 118:2466.

Martín, E. L., Barrado y Navascués, D., Baraffe, I., Bouy, H. and Dahm, S.(2003). A Hubble Space Telescope Wide Field Planetary Camera 2 Survey for Brown Dwarf Binaries in the $\alpha$ Persei and Pleiades Open Clusters. *ApJ*, 594:525.

Masciadri, E., Brandner, W., Bouy, H., Lenzen, R., Lagrange, A. M., and Lacombe, F. (2003). First NACO Observations of the Brown Dwarf LHS 2397aB. *A&A*, 411:157.

McCarthy, C. and Zuckerman, B. (2004). The Brown Dwarf Desert at 75-1200 AU. *AJ*, 127:2871.

McCaughrean, M. J., Close, L. M., Scholz, R. D. et al. (2004). $\epsilon$ Indi Ba,Bb: The Nearest Binary Brown Dwarf. *A&A*, 413:1029.

Moraux, E., Bouvier, J., Stauffer, J. R., Cuillandre, J.-C. (2003). Brown Dwarfs in the Pleiades Cluster: Clues to the Substellar Mass Function. *A&A*, 400:891.

Nagashima, C., Dobbie, P. D., Nagayama, T. et al. (2003). An Optical and Near-Infrared Search for Brown Dwarfs in the Pleiades Cluster. *MNRAS*, 343:1263.

Nakajima, T., Oppenheimer, B. R., Kulkarni, S. R., Golimowski, D. A., Matthews, K. and Durrance, S. T. (1995). Discovery of a Cool Brown Dwarf. *Nature*, 378:463.

Nakajima, T., Tsuji, T., and Yanagisawa, K. (2004). Spectral Classification and Effective Temperatures of L and T Dwarfs Based on Near-Infrared Spectra. *ApJ*, 607:499.

Pinfield, D. J., Dobbie, P. D., Jameson, R. F., Steele, I. A., Jones, H. R. A.and Katsiyannis, A. C. (2003). Brown Dwarfs and Low-Mass Stars in the Pleiades and Praesepe: Membership and Binarity. *MNRAS*, 342:1241.

Rebolo, R., Martín E. L., and Magazzù, A. (1992). Spectroscopy of a Brown Dwarf Candidate in the $\alpha$ Persei Open cluster . *ApJ* 389:L83.

Rebolo, R., Zapatero Osorio M. R., and Martín , E. L. (1995). Discovery of a Brown Dwarf in the Pleiades Star Cluster. *Nature*, 377:129.

Rebolo, R., Martín E. L, Basri, G., Marcy, G., and Zapatero Osorio, M., R. (1996). Brown Dwarfs in the Pleiades Cluster Confirmed by the Lithium Test. *ApJ* 469:L53.

Rebolo, R., Zapatero Osorio M. R., Madruga, S., Béjar V. J.S ., Arribas, S., and Licandro, J.(1998). Discovery of a Low-Mass Brown Dwarf Companion of the Young Nearby Star G 196-3. *Science*, 282:1309.

Reid, I. N., Kirkpatrick, J. D., Liebert, J. et al. (1999). L Dwarfs and the Substellar Mass Function *ApJ*, 521:613.

Reipurth, B. and Clarke, C. (2001). The Formation of Brown Dwarfs as Ejected Stellar Embryos *AJ*, 122:432.

Rieke, G. H., and Rieke M. J. (1990). Possible substellar objects in the $\rho$ Ophiuchi cloud. *ApJ* 362:L21.

Ruiz, M. T., Leggett, S. K., and Allard, F. (1997). Kelu-1: A Free-floating Brown Dwarf in the Solar Neighborhood. *ApJ* 491:L107.

Scholz, R. D., McCaughrean, M. J., Lodieu, N. and Kuhlbrodt, B. (2003). $\epsilon$ Indi B: A New Benchmark T dwarf. *A&A*, 398:L29.

Stauffer, J., Hamilton, D., Probst, R., Rieke, G. and Mateo, M. (1989). Possible Pleiades Members with Mass of about 0.07 solar mass- Identification of Brown Dwarf Candidates of Known Age, Distance, and Metallicity. *ApJ* 344:L21.

Stauffer, J., Herter, T., Hamilton, D., Rieke, G. H., Rieke, M. J., Probst, R. and Forrest, W. (1991). Spectroscopy of Taurus Cloud Brown Dwarf Candidates. *ApJ* 367:L23.

Stauffer, J., Liebert, J., Giampapa, M., Macintosh, B., Reid, N. and Hamilton, D. (1994). Radial Velocities of Very Low Mass Stars and Candidate Brown Dwarf Members of the Hyades and Pleiades. *AJ*, 108:160.

Whitworth, A. P. and Zinnecker, H.(2004). The Formation of Free-floating Brown Dwarves and Planetary-Mass Objects by Photo-Erosion of Prestellar Cores. *A&A*, 427:299.

Williams, D. M., Comerón, F., Rieke, G. H., and Rieke, M. J.(1995). The Low-Mass IMF in the $\rho$ Ophiuchi cluster. *ApJ*, 454:144.

Zapatero Osorio, M. R., Rebolo, R., Martín, E. L. et al.. (1997). New Brown Dwarfs in the Pleiades Cluster. *ApJ* 491:L81.

Zapatero Osorio, M. R., Béjar, V. J. S.., Martín, E. L., Rebolo, R., Barrado y Navascués, D., Bailer-Jones, C. A. L. and Mundt, R. (2000). Discovery of Young, Isolated Planetary Mass Objects in the $\sigma$ Orioniis Star Cluster. *Science*, 290:103.

Zapatero Osorio, M. R., Béjar, V. J. S.., Martín, E. L., Rebolo, R., Barrado y Navascués,D., Mundt, R., Eisloffel, J. and Caballero, J. A. (2002). A Methane, Isolated Planetary-Mass Object in Orion. *ApJ* 578:L536.

Zapatero Osorio, M. R., Lane, B. F., Pavlenko, Ya., Martín, E. L., Britton, M. and Kulkarni, S. R.(2004). Dynamical Masses of the Binary Brown Dwarf GJ 569 Bab. *ApJ*, 615:958.

Zuckerman, B. and Becklin, E. E. (1992). Companions to White Dwarfs: Very Low-mass Stars and the Brown Dwarf Candidate GD 165B. *ApJ*, 386:260.

II

# CONTRIBUTED POSTER ARTICLES

# REVEALING THE STRUCTURE OF LUPINE DARKNESS: FROM CORES TO CLUSTERS


Paula S. Teixeira[1,2,3], Charles J. Lada[1], and João Alves[4]

[1] *Smithsonian Astrophysical Observatory, Cambridge, MA, USA*

[2] *Faculdade de Ciências da Universidade de Lisboa, Portugal*

[3] *Centro de Física Nuclear da Universidade de Lisboa, Portugal*

[4] *European Southern Observatory, Garching, Germany*



**Abstract** We report in this paper results obtained from the analysis of deep near-infrared (NIR) observations of the densest region of the Lupus 3 cloud acquired from the European Southern Observatory's (ESO) New Technology Telescope (NTT) and Very Large Telescope (VLT). By constructing a dust extinction map (using the near-infrared color excess method) and examining the detailed internal structure of this prime example of a filamentary dark cloud, we have isolated 5 prominent embedded cores, a dense filament and a dense ring structure in this cloud. The cores have distinctive profiles, sizes, masses, some of them have on-going star formation while others are apparently starless. Building azimuthal density profiles and using Bonnor-Ebert and Jeans analysis, we are able to place these cores in an evolutionary sequence. We discuss the relation between the dense ring structure and the emerging cluster in the cloud. Assuming that the ring is the remnant of the disrupted core from which the small cluster originated, we calculate a corresponding star formation efficiency of $\approx 30\%$ for the original core. Lupus 3 is a very interesting cloud for having so many different structures within - where we identify both starless, stellar cores as well as a cluster emerging from a disrupted core. A correlation between the density structures of the cores and its stability and star formation status is evident.


## Introduction

In order to better understand the process of star formation it is essential to know what the initial conditions are. This is the key motivation for probing the structures of clouds. The Lupus 3 cloud presents itself as an ideal object to analyze for it is one of the nearest regions (140 pc Hughes & Hartigan 1993) with active star formation (28 PMS sources identified in H$\alpha$ by Schwartz 1997 in the region we survey). This paper presents results from dust extinction maps of

Lupus 3, from which we obtain two of the initial conditions for star formation: the density structure and the mass of the cloud.

## Observations and Data Reduction

Observations of the Lupus filamentary cloud were carried out in March 2000 using the SofI camera at European Southern Observatory's (ESO) New Technology Telescope (NTT) in La Silla, Chile. Figure 1 shows the observational layout: twelve frames were taken both in the $H$–band and $K_s$–band filters and 3 of those frames were also observed with the $J$–band filter to better analyze star formation in those regions. For two regions that are heavily extincted, and as such were not penetrated by the deep NTT observations, further observations were taken using ESO's Very Large Telescope (VLT) in Paranal, Chile. The dithered images were reduced with DIMSUM (Deep Infrared Mosaicing Software) and photometry was then performed using the FLAMINGOS PinkPhot pipeline. Photometric and astrometric calibrations were then performed using 2MASS data.

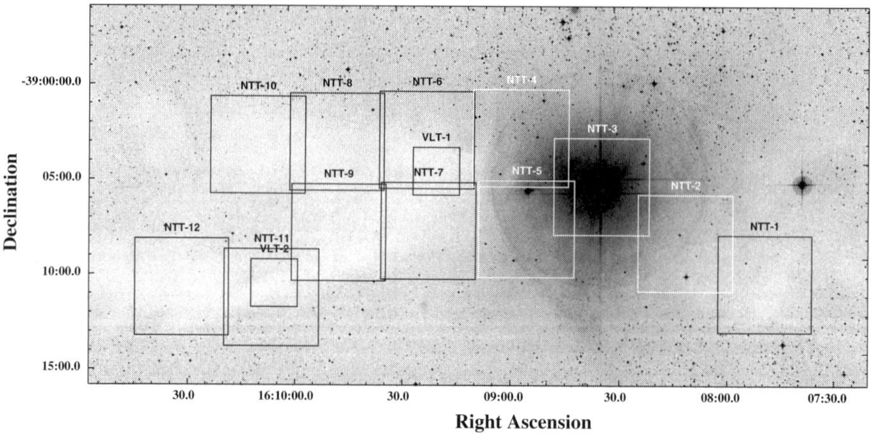


*Figure 1.* Optical image of the Lupus 3 molecular cloud obtained from the Digitized Sky Survey, blue POSS2. Our NTT surveyed area is marked by the bigger boxes while the two regions imaged with VLT are shown in the smaller boxes.

## Results

We use the near-infrared color excess (NICE) method described in Lada & Lada (1994) to derive extinctions for stars background to the Lupus 3 cloud. The color excess is given by subtracting the average $(H - K_s)$ color of the stars in a control field (assumed to be the intrinsic color of the background stars to cloud) from the $(H - K_s)$ of each star from the field on the cloud. Our control field is located 18′ north of the Lupus 3 cluster and has $<(H - K_s)> = 0.19 \pm 0.11$ magnitudes. To convert color excesses to equivalent visual extinctions, we adopt a reddening law from Rieke & Lebofsky (1985), $A_v = 15.93\,E(H - K_s)$.

Figure 2 shows the dust extinction map we obtained for the Lupus 3 complex using 2MASS $H-$ and $K_s-$ band data and the NICE method mentioned above. This map shows a very dense central part and lower density filamentary

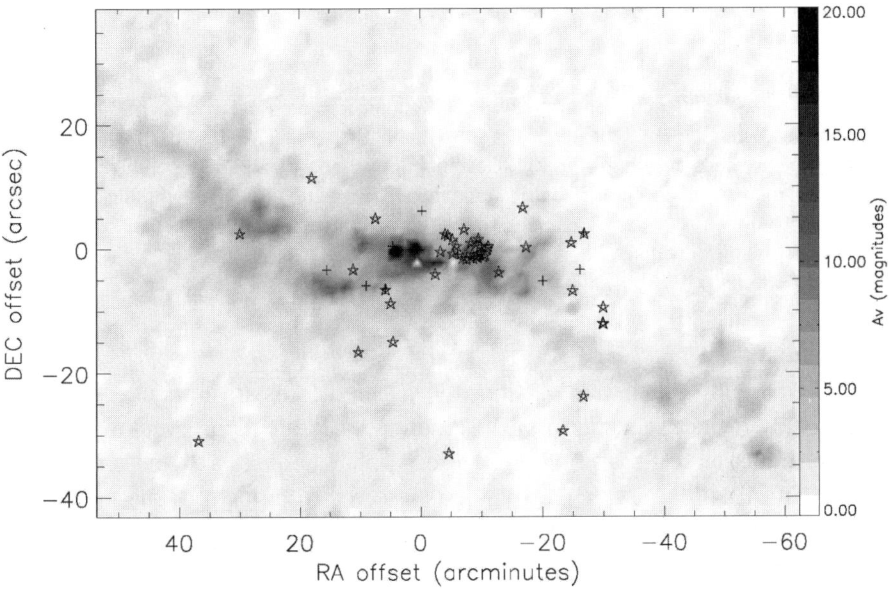


*Figure 2.* Nyquist sampled, 2′ resolution dust extinction map of the Lupus 3 cloud complex built using 2MASS data. The (0,0) position is ($16^h 09^m 18.8^s$,$-39^o 04' 48.7''$)(J2000). The stars mark the positions of H$\alpha$ stars from Schwartz(1977) and the crosses represent catalogued IRAS sources.

structures that extend from it. We have built a more detailed dust map of the denser central filamentary cloud using the data obtained at the NTT and the VLT, shown in Figure 3. The more detailed map has allowed us to identify dense embedded cores (labeled from A to H), a dense filament and a dusty ring very close to the small cluster in the Lupus 3 cloud.

We have calculated masses from the integration of the dust column density from the dust extinction map (for details see Cambrésy, 1994) for the whole cloud surveyed, as well as for five prominent cores (A, B, C, D, and H). The total mass calculated for Lupus 3 is $84.7 \pm 0.8\,M_\odot$, while the cores have total masses ranging between $12.2\,M_\odot$ to $1.3\,M_\odot$. Finally, the ring structure has an

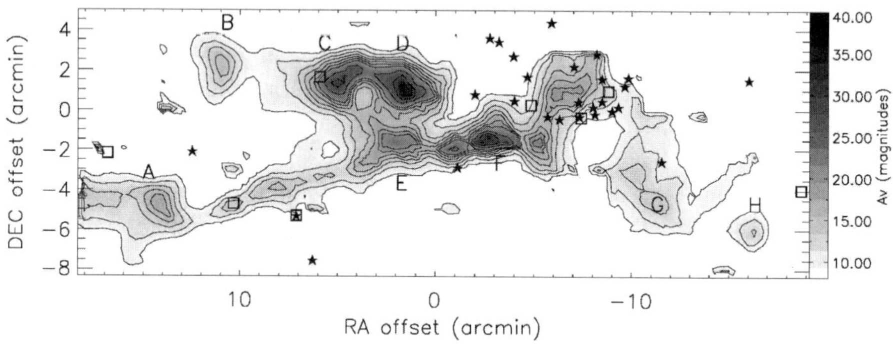


*Figure 3.* Dust extinction map of the Lupus 3 cloud. The beam is a Gaussian of 30″ FWHM, Nyquist sampled. Cores and filaments discussed in the text are labeled. The (0,0) position is the same as the previous dust map (fig. 2). The extinction contours start at 8 magnitudes of visual extinction, increasing in steps of 2 until 20 and then in steps of 5 until 40 magnitudes. The stars mark the positions of H$\alpha$ stars from Schwartz(1977) and the crosses represent catalogued IRAS sources (see text for details).

estimated total mass of 25 $M_\odot$. Since the selected cores are embedded we have subtracted off the contribution of the cloud to calculate their actual mass. This is done by assuming that the cores lie on a plateau of approximately 6 magnitudes of visual extinction (this value is inferred from the dust map and from individual radial profiles that will be discussed below). The final corrected and adjusted masses for the cores range between 8.5 $M_\odot$ to 0.9 $M_\odot$.

We have built radial extinction profiles for each core and fitted the empirical profiles we obtained with Bonnor-Ebert sphere profiles (for discussion on BE spheres see e.g. Lada et al., 2004 and Harvey et al., 2003). From this fitting we have estimated physical parameters, such as central density, center-to-edge density contrast ($\rho_c/\rho_e$ ($\rho_e = \rho(R)$), external pressure, and mass. These are summarized in Table 1. The Bonnor-Ebert mass cited is in very good agreement with the mass we derive from the dust extinction map.

## Analysis and Discussion

Comparison of the physical parameters estimated using Bonnor-Ebert spheres suggests that there exists a relation between the evolutionary status of the cores and their density structure. The cores with highest center-to-edge density contrast correspond to higher values of $\xi_{\rm max}$. For true Bonnor-Ebert equilibrium configurations, $\xi_{\rm max}$ may be related to the evolutionary state of a Bonnor-Ebert cloud. High values of $\xi_{\rm max}$ and steeply rising density profiles would presumably correspond to unstable regions undergoing star formation, while small

*Table 1.* Derived properties from Bonnor-Ebert fitting.

| *Core* | $A_V(max)^a$ *(magnitudes)* | *R* *(pc)* | $\xi_{max}$ | $T_{BE}$ *(K)* | $n_c$ $(10^4 cm^{-3})$ | $P_{ext}/k$ $(10^4\ cm^{-3} K)$ | $M_{BE}$ $(M_\odot)$ |
|---|---|---|---|---|---|---|---|
| A | 16.3 | 0.07 | 6.7 | 9.5 | 10.6 | 5.8 | 1.3 |
| B | 12.2 | 0.07 | 5.8 | 9.0 | 5.9 | 5.0 | 1.2 |
| C | 33.5 | 0.09 | 8.7 | 18.5 | 20.7 | 10.9 | 3.3 |
| D | 40.2 | 0.11 | 7.5 | 30.6 | 18.8 | 29.0 | 6.4 |
| H | 14.7 | 0.06 | 10.6 | 4.5 | 17.2 | 1.5 | 0.5 |

[a] Maximum visual extinction given by the Bonnor-Ebert fit

values of $\xi_{\max}$ and relatively shallow density gradients would correspond to stable or perhaps pre-star forming cores.

An independant assessment of the stability of these cores can be obtained by determining their Jeans masses. From this evaluation, we conclude that cores A and B are Jeans stable, while cores C and D are Jeans unstable. This agrees with our conclusions regarding the stability of the cores using the Bonnor-Ebert criterion analysis. Both these methods assume only thermal support against gravitational collapse.

Further examination of the evolutionary status of the Lupus 3 cores requires a census of star forming activity within the cloud. To pursue this we have searched the literature as well as existing IRAS and *Spitzer* data. Cores A and B have no IRAS sources catalogued, and inspection of the *Spitzer* images from the data publicly available from the Cores to Disks *Spitzer* Legacy Program (Evans et al. 2003) shows no MIPS sources detected, hence we conclude that cores A and B are likely starless. There is an IRAS source located in the filament, IRAS 16067-3902, (see Figure 3) and in core C, IRAS 16063-3856, however, these sources have only upper limits to fluxes in all the bands and inspections of the IRAS images do not show convincing proof for point source detection so these detections could be spurious. We therefore assume core C, as well as the filament that connects core A to core E, are starless. There is evidence for star formation in core D: the Herbig-Haro object HH 78 (Reipurth & Graham 1988). We clearly see in out VLT $H-$ and $K_s-$ band images of core D an elongated source in the southeast which we believe is evidence of a jet associated with HH78 and nebulosity in the northeast that would correspond to a cavity around a counter-jet. There is an IRAS source catalogued for this core, IRAS 16059-3857, however these detections are marginal.

Overall we see a progression of core evolution from starless stable cores, to unstable star forming cores, and finally, to an emerging cluster which has disrupted its parental core.

## Acknowledgments

Support for this work was provided by NASA through grants NAG5-9520 and NAG5-13041. P. Teixeira acknowledges support from the Fundação para a Ciência e Tecnologia (FCT) Programa Operacional Ciência Tecnologia Inovação (POCTI) do Quadro Comunitário de Apoio III, graduate fellowship SFRH/BD/13984/2003, Portugal.

## References

Alves, J. F., Lada, C. J. & Lada, E. A 2001, Nature, 409, 159

Andreazza, C. M. & Vilas-Boas, J. W. S. 1996, A&AS, 116, 21

Bohlin, R. C., Savage, B. D. & Drake, J. F. 1978, ApJ, 224, 132

Bonnor, W. 1956, MNRAS, 116, 351

Cambrésy, L. 1999, A&A, 345, 965

Ebert, R. 1955, Z. Astrophys., 37, 217

Evans, N. J., et al. 2003, PASP, 115, 965

Hara, A., Tachihara, K., Mizuno, A., Onishi, T., Kawamura, A., Obayashi, A & Fukui, Y. 1999, PASJ, 51, 895

Harvey, D. W. A., Wilner, D. J., Lada, C. J., Myers, P. C., & Alves, J. F. 2003, ApJ, 598, 1112

Lada, C. J., Lada, E. A., Clemens, D. & Bally, J. 1994, ApJ, 429, 694

Lada, C. J. & Lada, E. A. 2003, ARA&A, 41, 57

Lada, C. J., Huard, T. L., Crews, L. J., & Alves, J. F. 2004, ApJ, 610, 303

Nakajima, Y., Tamura, M., Oasa, Y. & Nakajima, T. 2000, AJ, 119, 873

Rieke, M. J. & Lebofsky, L. A. 1985, ApJ, 288, 618

Reipurth, B. & Graham, J. A. 1985, ApJ, 288, 618

Schwartz, R. 1977, ApJS, 35, 161

Tachihara, K., Onishi, T., Mizuno, A. & Fukui, Y. 2002, 385, 909

Vilas-Boas, J. W. S., Meyers, P. C. & Fuller, G. A. 2000, 532, 1038

# INFRARED STUDY OF THE SOUTHERN GALACTIC STAR FORMING REGION ASSOCIATED WITH IRAS 10049-5657

S. Vig, S. K. Ghosh, D. K. Ojha, R. P. Verma
*Tata Institute of Fundamental Research, Mumbai, 400 005, India*

**Abstract** The southern Galactic star forming region associated with IRAS 10049-5657 has been mapped simultaneously in two far infrared bands ($\lambda_{eff}$ = 150 & 210 $\mu$m), with $\sim 1'$ angular resolution using the TIFR 1-m balloon borne telescope. Spatial distribution of the temperature of cool dust and optical depth at 150 $\mu$m, have been obtained taking advantage of the similar beams in the two bands. The HIRES processed IRAS maps at 12, 25, 60 & 100 $\mu$m have been used for comparison. The distribution of warm dust emission as well as emission in the Unidentified Infrared Bands in the mid infrared have been studied, based on the MSX data (8, 12, 14, & 21 $\mu$m bands). Radiative transfer modelling has been carried out taking into account all available infrared measurements to obtain various physical parameters like nature of the exciting source, geometrical dimensions, dust distribution and composition etc.

## Introduction

HII regions are astronomical sources that represent early stages of deeply embedded high mass (O or early B) stars. Their study can provide vital information about the formation of massive stars as well as their interaction with the parent molecular cloud. Being deeply embedded in dust, almost all of their energy is emitted in the infrared waveband. In this paper, we report far and mid infrared observations of the southern Galactic star forming region IRAS 10049-5657 (extended HII region). Limited information exists about this star forming region. It is at a distance of $\sim 5.1$ kpc (Vilas-Boas & Abraham, 2000). CS line emission and $H_2CO$ emission (Whiteoak & Gardner, 1974) have been detected close to this source.

## Observations

IRAS 10049-5657 has been observed using the 12-channel two-band far-infrared (FIR) photometer system at the Cassegrain focus of the TIFR 100 cm (f/8) balloon borne telescope. The observations were carried out during the balloon flight from the TIFR Balloon Facility, Hyderabad on February 20, 1994. The angular resolution of the deconvolved maps is $\sim 1'$.

In addition, the spatial distribution of emission in Unidentified Infrared Bands (UIBs) has also been computed for this source using the technique of Ghosh & Ojha (2002) from Midcourse Space Experiment (MSX) data (Price et al, 2001) using four mid infrared bands (8, 12, 14 & 21 $\mu$m).

IRAS survey data for this source was processed using the HIRES routine at the Infrared Processing and Analysis Center (IPAC) to obtain high resolution images in all the four IRAS bands at 12, 25, 60 and 100 $\mu$m. Details of the HIRES routine have been given in Aumann et al. (1990). The IRAS-LRS spectrum of this source is also used for construction of the spectral energy distribution (SED) in the present work.

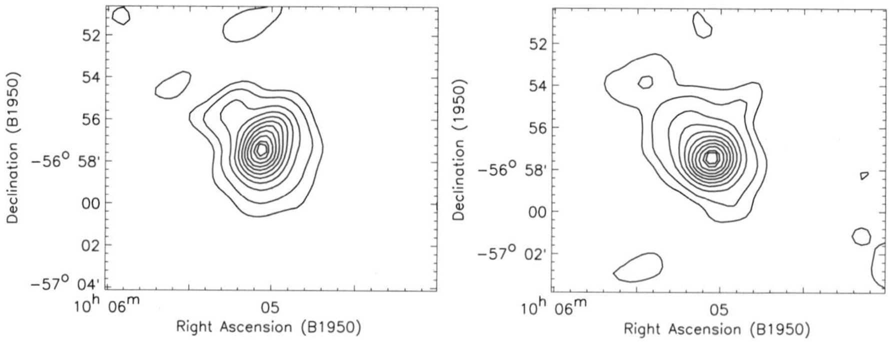


*Figure 1.* The intensity map for the region around IRAS 10049-5657 in (left) 150 $\mu$m, and (right) 210 $\mu$m bands (1950 coordinates). Contour levels are at 5, 10, 20, 30 40, 50, 60, 70, 80, 90 and 95% of peak intensity of (left) 624 Jy/sq arc min and (right) 292 Jy/sq arc min, respectively.

## Results and discussion

An area of $\sim 32' \times 20'$ was scanned around IRAS 10049-5657. The observed sky chopped far infrared signals have been deconvolved using an indigenously developed procedure (Ghosh et al., 1988) based on Maximum Entropy Method (MEM). Taking advantage of the simutaneous observations in the two bands, with almost identical field of view, we have generated maps of the dust temperature ($T_{\rm d}$) and optical depth at 150 $\mu$m ($\tau_{150}$). For these maps we have assumed dust emissivity of the form $\epsilon_\lambda \propto \lambda^{-2}$.

The deconvolved maps of IRAS 10049-5657 at 150 and 210 $\mu$m are presented in Figure 1. The far infrared emission samples the cold dust around these regions. We have detected dust as cold as 20 K. The peak optical depth at 150 $\mu$m around this region is determined to be $1.2 \times 10^{-3}$. The dust temperature T(150/210) and optical depth ($\tau_{150}$) maps are shown in Figure 2. From the dust emission around these regions, we see an extension of the dust emission towards north-east.

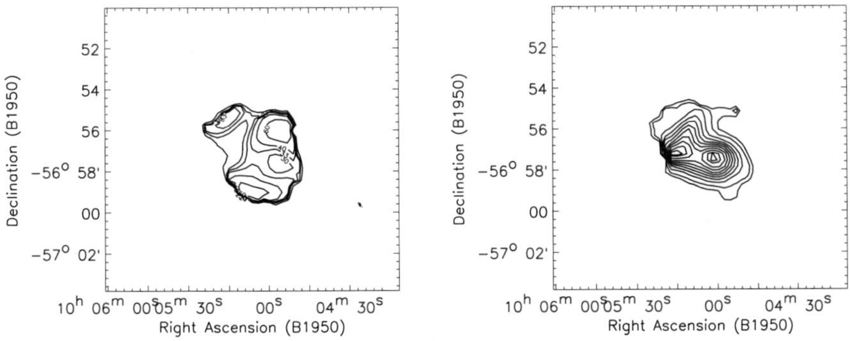


*Figure 2.* (Left) Dust temperature T(150/210) and (right) optical depth ($\tau_{150}$) distribution for the region around IRAS 10049-5657. Contours are at the levels of (left) 20, 27, 30, 33, 40 & 60 K and (right) at 5, 10, 20, 30, 40, 50, 60, 70, 80, 90 and 95% of the peak value which is $1.2 \times 10^{-3}$.

In Figure 3, we have plotted the contours of the region due to the extracted emission in UIBs in the $6 - 9$ $\mu$m region. A comparison between thermal emission at 8 $\mu$m and distribution in UIBs indicates that the spatial distribution of the warm dust emission is quite similar to the emission in UIBs.

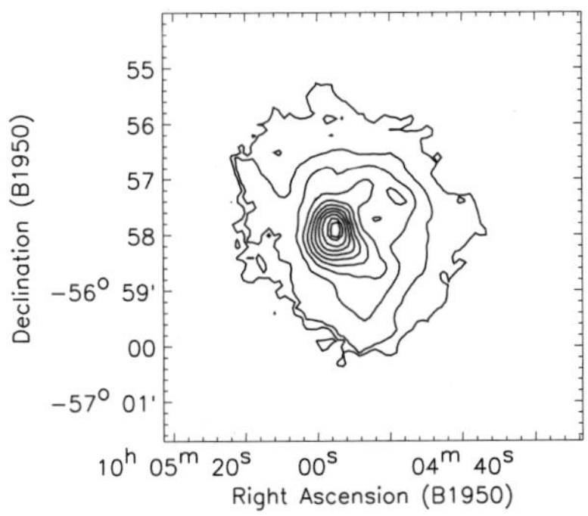


*Figure 3.* Distribution of modelled emission in Unidentified Infrared Bands (UIBs) for the region around IRAS 10049-5657. Contours are at the levels of 5, 10, 20, 30, 40, 50, 60, 70, 80, 90, 95% of peak intensity of $8.9 \times 10^{-5}$ W m$^{-2}$ Sr$^{-1}$.

The IRAS bands of 12, 25, 60 and 100 $\mu$m probes dust warmer than that probed by the 150 and 210 $\mu$m bands, around this region. Figure 4 shows the spatial morphology of the warm dust emission in all the four IRAS bands.

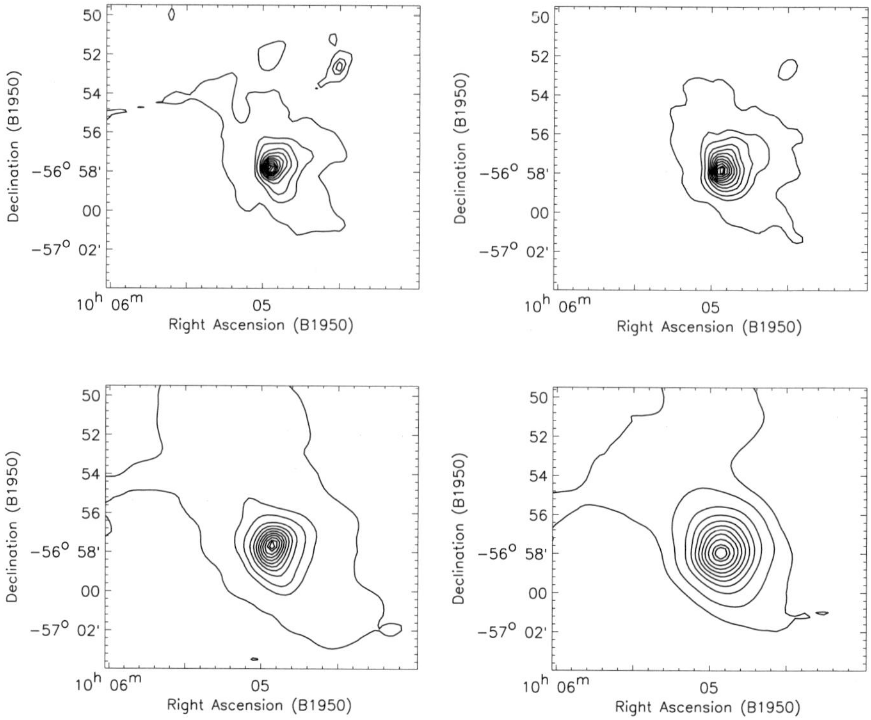


*Figure 4.* The HIRES processed intensity map of the region around IRAS 10049-5657 in (top left) 12 $\mu$m, (top right) 25 $\mu$m, (bottom left) 60 $\mu$m and (bottom right) 100 $\mu$m (1950 coordinates). The contour levels are at 1, 5, 10, 20, 30, 40, 50, 60, 70, 80, 90, 95% of the peak. The peak intensities are 272, 1053, 3367 and 1423 Jy/sq arc min at 12, 25, 60 and 100 $\mu$m, respectively.

## Radiative transfer modelling

The star forming region is modelled as a spherically symmetric cloud of gas (hydrogen) and dust, powered by a centrally embedded source (single or cluster of ZAMS stars). The cloud is immersed in an average interstellar radiation field. The observational constraints include the SED due to dust and angular sizes at different wavelengths. Other details of this self-consistent scheme are given in Mookerjea & Ghosh (1999). The parameters explored are the nature of embedded source, radial density distrbution laws ($n(r) \sim r^0, r^{-1}, r^{-2}$), relative abundances of different grain types (silicate, graphite), radial optical depth due to dust, gas-to-dust ratio and the geometric dimensions of the cloud. Two commonly used types of interstellar dust are explored, the DL type (Draine & Lee, 1984) and the MMP type (Mathis, Mezger & Panagia, 1983).

The SED for IRAS 10049-5657 is constructed using the flux densities at the four MSX bands, the IRAS-LRS data, the four IRAS bands (from HIRES maps) and the two TIFR bands. The flux densities used in the SED are the fluxes integrated in a circular region of diameter $3'$ around the peak. The total luminosity is $6.5 \times 10^5$ $L_\odot$ for a distance of 5.1 kpc. Figure 5 shows the observed and the best fit model of the SED. The best fit model of radiative transfer leads to a uniform dust and gas density distribution with the embedded energy source as a single ZAMS star of type O5. The inner cloud dust radius is 0.008 pc while the outer cloud radius is 3.4 pc. The optical depth at 100 $\mu$m is 0.002. The radius of the ionised gas from the model is 2.6 pc for gas-to-dust ration of 100:1 by mass. The dust composition for Si:C is 11:89 for the DL type of dust.

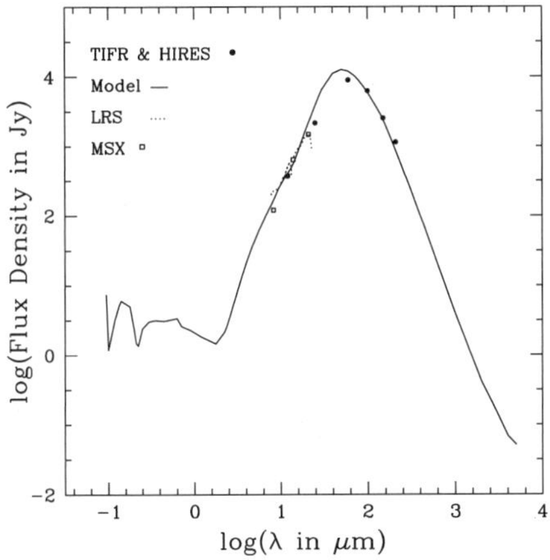


*Figure 5.* Comparison of the predicted spectral energy distributions (SED) from the best fit models of IRAS 10049-5657. The filled circles represent the TIFR and IRAS-HIRES data. The dashed line is the IRAS-LRS spectra. The open squares represent the MSX data. The solid line denotes the best fit model to the data. See text for details.

## Summary

The southern Galactic star forming region associated with IRAS 10049-5657 has been mapped in two far infrared wavebands at 150 and 210 $\mu$m with an angular resolution of $\sim 1'$. Reliable dust color temperature and optical depth (at 150 $\mu$m) maps have also been obtained. The IRAS-HIRES data for this region has also been used for comparison. Self-consistent radiative transfer calculations have been carried out through a spherical gas-dust cloud. A radial distribution of $r^0$,i.e. constant density distribution, is preferred. The geometric details of the gas-dust clouds, the dust composition and optical depth etc. have been quantified by the model.

## References

Aumann, H. H., Fowler, J. W. & Melnyk, M. 1990, *Astron. Jour.*, **99**, 1674
Draine, B. T. & Lee, M. H. 1984, *Astrophys. J.* **285**, 89
Ghosh, S. K., Iyengar, K. V. K., Rengarajan *et al.* 1988, *Astrophys. J.*, **330**, 928
Ghosh, S. K., & Ojha, D. K. 2002, *Astron. & Astrophys.*, **388**, 326
Mathis, J. S., Mezger, P. G. & Panagia, N. 1983, *Astron. & Astrophys.*, **128**, 212
Mookerjea, B. & Ghosh, S. K., 1999, *J. Astrophys. Astr.*, **20**, 1
Price, S. D., Egan, M. P., Carey *et al.* 2001, *Astrophys. J.*, **121**, 2819
Vilas-Boas, J. W. S., & Abraham, Z. 2000, *Astron. & Astrophys.*, **355**, 115
Whiteoak, J. B. & Gardner, F. F. 1974, *Astron. & Astrophys.*, **37**, 389

# STUDYING PROTOSTELLAR JETS THROUGH A COMBINED INFRARED/OPTICAL SPECTRAL ANALYSIS

L. Podio[1], F. Bacciotti[2], B. Nisini[3], T. Giannini[3], F. Massi[2], J. Eislöffel[4], T.P. Ray[5]

[1]*Dipartimento di Astronomia e Scienza dello Spazio, Università di Firenze, L.go E. Fermi 2, 50125 Firenze, Italy, lindapod@arcetri.astro.it*

[2]*INAF-Osservatorio Astrofisico di Arcetri, L.go E. Fermi 5, 50125 Firenze, Italy*

[3]*INAF-Osservatorio Astronomico di Roma, Via di Frascati 33, 00040 Monte Porzio Catone, Italy*

[4]*Thüringer Landessternwarte Tautenburg, Sternwarte 5, D-07778 Tautenburg, Germany*

[5]*School of Cosmic Physics, Dublin Institute for Advanced Studies, 5 Merrion Square, Dublin 2, Ireland*

**Abstract** Supersonic protostellar jets are an essential element to understand obscure aspects of the star formation process. To investigate their physical structure and to prepare for the observations with high angular resolution, we have studied the conditions of the plasma along a sample of 'classical' Herbig-Haro (HH) jets through a combined optical-infrared diagnostics. Here we present and discuss the results obtained for the HH 1 and HH 34 jets. We applied our diagnostic tecnique to moderate spatial/spectral resolution data (3.6m-EFOSC2 and NTT-SOFI) that cover a wide range from 0.6 - 2.5 micron. >From the emission line fluxes we have derived parameters such as the variation of the visual extinction along the flow, the electron density, the ionization fraction, the temperature, the total density, and the mass flux rate. The latter is an important parameter to understand the role of protostellar jets in the star formation process (removal of excess angular momentum, dispersion of infalling enevlope, injection of turbolence for cloud support). Moreover, we investigated the abundances of refractory emitting species such as Carbon, Calcium and Iron. The depletion of these species with respect to the Solar values allows us to understand if there are still dust grains along the flow. In the future the new generation instruments being developed for high angular resolution and interferometry (VLTI/AMBER and LBT) will allow us to investigate the gas conditions at smaller scales, where the acceleration and collimation of the jet take place.

## Introduction

Jets are believed to be an essential ingredient of the star formation process (e.g. Königl & Pudritz 2000; Reipurth 2001), and thus is is important to investigate their properties with dedicated observations. The theories of jet propagation predict that density and temperature stratifications are generated in the beam, due to the action of the compressing shocks (Hartigan et al. 1994). For example the optical [SII], [NII] and [OI] lines originate in a region of the post-shocked gas near to the shock-front, at intermediate temperatures, $T_e$, and ionization fractions, $x_e$, while [FeII] transitions are excited in a further and more compressed region, at higher density and lower temperature. Such a stratification can be tested with spectroscopic observations in the optical/NIR range, that comprises transitions from atomic, single ionized and molecular species which have different excitation temperatures and critical densities. To this aim we have taken infrared and optical spectra of a sample of classical protostellar jets (HH 1, HH 34, HH 83, HH 73, HH 111, HH 24C/E, HH 24G, VELA-irs8) with 3.6m-EFOSC2 and NTT-SOFI (R $\sim$ 600). Through the observed emission lines we have determined important parameters like the electron and total density, the ionization fraction and the temperature (Sect.1). >From the lines of refractory species like Calcium, Carbon and Iron, we derived information about the abundances of these species and thus the presence of dust in the flow (Sect. 2). Finally we inferred the mass flux through the jets (Sect. 3).

## Variation of the physical parameters along the jet

We applyed our combined optical/NIR diagnostics to the different knots identified along the jets, and analysed the variation of the physical parameters in the flow. The infrared ratios [FeII]1.64/1.25$\mu$m and [FeII]1.64/1.32$\mu$m allow us to estimate the visual extinction $A_V$ along the jet and therefore to use also ratios between lines distant in wavelength as a diagnostic instrument. >From the optical [SII], [NII] and [OI] lines, using the so called BE tecnique (Bacciotti & Eislöffel 1999), we determine the physical conditions of the gas ($T_e$, $x_e$, the electron and total density $n_e$ and $n_H$) in the region of optical emission (see Fig.1). For the HH 34 and HH 1 jets we inferred values of the electron density of $10^2 \div 4 \cdot 10^3$ cm$^{-3}$, ionization fractions of $0.03 \div 0.3$, $T_e$ between 10000 and 15000 K for HH 34 and up to 20000 K for HH 1. We derived from the obtained $x_e$ and $n_e$ total densities of $10^3 \div 3 \cdot 10^4$ cm$^{-3}$ in HH 34, and up to $10^5$ cm$^{-3}$ in HH 1. We than derived in an independent way the physical parameters in the region of Iron emission. >From the ratios [FeII]1.64/1.60$\mu$m and [FeII]1.64/1.53$\mu$m we infer the electron density, while the temperature is found using both optical and infrared [FeII] lines (from [FeII]1.64/0.86$\mu$m, [FeII]1.64/0.90$\mu$m, [FeII]1.64/0.92$\mu$m, [FeII]1.64/0.93$\mu$m (Nisini et al. 2002; Pesenti et al. 2003)). As shown in

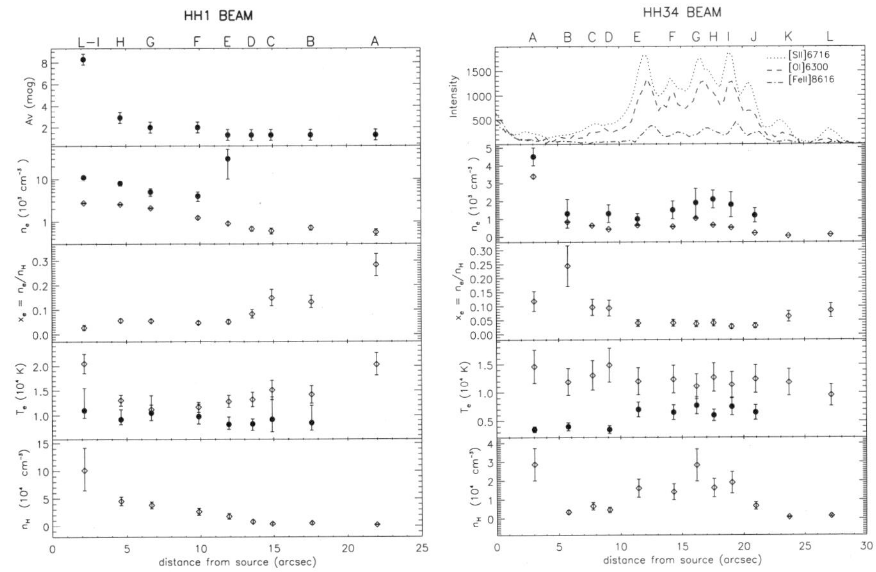


*Figure 1.* Variations of the physical parameters along the HH 1 and HH 34 jets. The diamonds are the values derived from optical $S^+$, $O^0$, $N^+$ lines. Circles indicate the values derived from [FeII] lines, that trace the denser and cooler portion of the post-shocked gas.

Fig.1, from [FeII] lines we found values of $n_e$ higher than those derived from the optical lines ($n_e$([FeII]) $\sim 10^3 \div 10^4$ cm$^{-3}$), while the values of $T_e$ are lower ($T_e$ ([FeII]) $< 10^4$ K). These results agree with the fact that [FeII] lines trace a component of the post-shocked gas denser and cooler with respect to the component traced by the optical lines (Nisini et al. 2005, in prep.). Finally, in order to investigate if there is an even denser component inside the jet beam, we use the dereddened ratios [FeII]$\lambda$7155/8617 (Hartigan et al. 2004; Bautista & Pradhan 1998) and CaII$\lambda$8542/[CaII]$\lambda$7290, that are good tracers of very high density gas. We found values of $n_e \sim 10^5 \div 10^6$ cm$^{-3}$ both in HH 1 and HH 34, testifying the presence of a high density "spine" along these jets.

## Depletion of refractory species and presence of dust in the jet beam

According to the current knowledge of interstellar dust properties and structure, we expect that dust grains are completely destroyed by jet shocks and that the atoms of refractory species are all released in gaseous form. Since in our spectra we detect emission lines from these species, we can check if their abundances are close to the solar values, as expected. Comparing the observed and predicted emission of refractory and non-refractory species we found that Calcium, Carbon and Iron are depleted with respect to the Solar abundances, both along the HH 1 and the HH 34 jet (see Fig.2). This implies that some atoms of these species are still locked on dust grains and that the weak shocks along the beam are not able to completely destroy the dust. Since the properties of jet shocks are very well known, results of this kind are very useful to improve our knowledge of dust structure. This is fundamental both for observative applications (visual extinction, light polarisation) and for the study of many processes, including cloud chemistry, the evolution of disk and the formation of stars and planets.

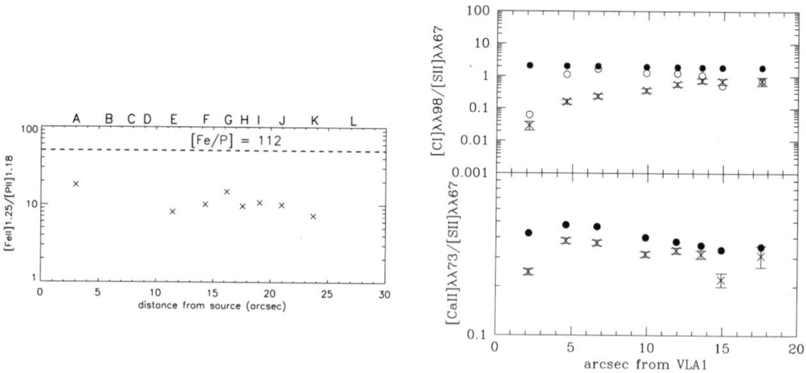


*Figure 2.* Observed and predicted ratios between lines of refractory (Ca, C, Fe) and non-refractory (S, P) species. *Left*: The dotted line indicates the predicted [FeII]1.25$\mu$m/[PII]1.18$\mu$m ratio, computed assuming solar abundances and all Fe and P ionized; crosses are the dereddened ratios measured in the knots along the HH 34 jet. The distance between the crosses and the line suggests that there is a depletion of Iron. We obtain the same result along the HH 1 jet. *Right*: Crosses are the dereddened ratios in the knots along the HH 1 jet; filled circles are the predicted ratios assuming solar abundances and all Calcium in $Ca^+$ and all Carbon in $C^0$ forms; open circles refer to the predictions computed with $C^0/C^+$ fraction given by ionization equilibrium analysis. The discrepancy between the observed and the predicted ratios is attributed to C and Ca depletion. We find a depletion of these species also along the HH 34 jet.

## Mass flux estimates and their implications

The mass flux rate ($\dot{M}_{\rm jet}$) is a fundamental parameter to be determined because it governes all the jet dynamics, and allows us to estimate the linear and angular momentum carried by the jet ($\dot{P}_{\rm jet}$ and $\dot{L}_{\rm jet}$ respectively). Through these quantities one can illustrate the central role of protostellar jets in important aspects of the star formation process. Determining $\dot{P}_{\rm jet}$ we can check if the jet is powerful enough to accelerate the sorrounding molecular outflows, to clear the star's environment and inject turbulence in the cloud. $\dot{L}_{\rm jet}$ is related instead to the jet capability of removing the excess angular momentum from the disk/star system, thus allowing the accretion of matter from the disk onto the central star (Königl & Pudritz 2000; Woitas et al. 2005, in prep.). Moreover the mass flux is an important paramater for a comparison between observations and theoretical models. In fact, in all the magneto-hydro-dynamic models proposed to explain the jet formation and acceleration, the ratio between the rate of mass ejected into the jet ($\dot{M}_{\rm jet}$) and the rate of mass accreted from the disk onto the star ($\dot{M}_{\rm acc}$) is fixed ($\dot{M}_{\rm jet}/\dot{M}_{\rm acc} \sim 0.01 \div 0.1$).

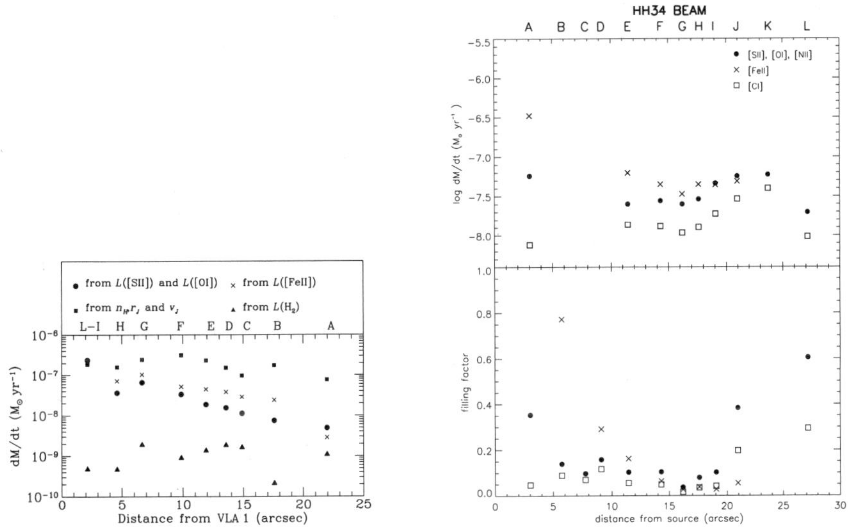


*Figure 3.* Mass flux rate along the HH 1 and the HH 34 jets. *Left*: In HH 1 $\dot{M}_{\rm jet}$ is computed with two methods: through the estimated total density (squares) and through the luminosity of different lines (circles are the values derived from [SII], [OI] line luminosity; crosses are values derived from [FeII] line luminosity assuming that all Fe is ionized). Triangles are the mass fluxes carried by the molecular component $H_2$. *Right*: $\dot{M}_{\rm jet}$ is computed through the luminosity of different lines. In the lower panel the derived filling factors in the knots are shown.

We determine the mass flux rate through the knots along the jet using two different methods (Fig.3). In the first method we estimate $\dot{M}_{\rm jet}$ from the total density derived through our diagnostics, and the jet radius $r_J$ and velocity $v_J$ derived from Hubble Space Telescope (HST) data (Reipurth et al. 2000). Using this method we implicitely assume that the knots are uniformely filled at the derived density, which gives un upper limit to $\dot{M}_{\rm jet}$. This effect is partially compensated by the fact that we are not taking into account the presence in the beam of the components denser than traced by [SII] lines. In the second method we estimate $\dot{M}_{\rm jet}$ from the dereddened flux of optically thin lines and we compare their absolute luminosity with the predicted emissivity for unit volume and unit total density (Hartigan et al. 1994). In this way we implicitely take into account the real filling factor in the beam, but our estimate is affected by the uncertanties in the flux calibration, visual extinction and distance. Overall we obtain $\dot{M}_{\rm jet} \sim 1 \cdot 10^{-7}$ M$_\odot$ yr$^{-1}$ for HH 1, and $\dot{M}_{\rm jet} \sim 5 \cdot 10^{-8}$ M$_\odot$ yr$^{-1}$ for HH 34. >From these results we see that the mass flux carried by the HH 34 jet is less than the one carried by HH 1. This agrees with theoretical expectations, because the HH 34 jet is ejected by a more evolved source, visible both in the IR and in the optical. For the HH 1 jet we estimated also $\dot{M}_{\rm jet}$ carried by the molecular component, using $H_2$ roto-vibrational transition lines. We obtain a mass flux $\sim 10^{-2} \div 10^{-3}$ times lower than that determined from optical lines (see Fig.3). This means that the molecular component has a negligible contribution in the jet excitation and dynamics with respect to the atomic component. >From these values of $\dot{M}_{\rm jet}$ we infer a flux of linear momentum of $\sim 3 \cdot 10^{-5} M_\odot$ yr$^{-1}$ km s$^{-1}$ for HH 1 and $\sim 1.5 \cdot 10^{-5} M_\odot$ yr$^{-1}$ km s$^{-1}$ for HH 34. These are of the same order of magnitude than those measured for the molecular outflows seen in CO lines. Thus the jets can indeed drive such flows. It would be very interesting to estimate the flux of angular momentum extracted by the jet, but for this we would need measurements of the jet toroidal velocity, which requires sub-arcsecond angular resolution observations (see, e.g., Bacciotti et al. 2002). To this aim we plan to examine existing HST archive data of these flows.

## References

Bacciotti, F., Eislöffel, J., 1999, A&A, 342, 717

Bacciotti, F., Ray, T.P., Mundt, R., Eislöffel, J., Solf, J., 2002, 576, 222

Bautista, M. A., Pradhan, A. K., 1998, ApJ, 492, 650B

Hartigan, P., Morse, J. A., Raymond, J., 1994, ApJ, 436, 125H

Hartigan, P., Edwards, F., Pierson, R., 2004, ApJ, 609, 261

Königl, A. & Pudritz, R., 2000, in Protostars and Planets IV, eds. V. Mannings, A. P. Boss, S. S. Russell (Tuscon: Univ. Arizona Press), 759

Nisini, B., Caratti o Garatti, A., Giannini, T., Lorenzetti, D., 2002, A&A, 393, 1035

Nisini, B., Bacciotti, F., Giannini, T., Massi, F., Eislöffel, J., Podio, L., Ray, T.P., 2005, A&A, in prep.

Pesenti, N., Dougados, C., Cabrit, S., O'Brien, D., Garcia, P., Ferreira, J., 2003, A&A, 410, 155

Reipurth, B., Heathcote, S., Yu, K. C., Bally, J., Rodriguez, L. F., 2000, ApJ, 534, 317R

Reipurth, B., 2001, Nature, 409, 140R

Woitas, J., Eislöffel, J., 2005, A&A, in prep.

# PRELIMINARY RESULTS ON NEWLY DISCOVERED EMBEDDED CLUSTERS

András Gáspár,[1] Zoltán Balog,[1] Zoltán Makai,[2] József Vinkó,[1] and Scott J. Kenyon[3]

[1] *Dept. of Optics and Quantum Electronics, University of Szeged, Szeged, Hungary*

[2] *Dept. of Experimental Physics, University of Szeged, Szeged, Hungary*

[3] *Smithsonian Astrophysical Observatory, Cambridge, MA USA*

gaspara@titan.physx.u-szeged.hu, balogz@titan.physx.u-szeged.hu

**Abstract** We present the preliminary results of an IR survey targeting newly discovered embedded clusters. The clusters were discovered by Bica and Dutra (2003) during a systematic search of the 2MASS database. The aim of our survey was to reobserve these objects with better spatial resolution and limiting magnitude. To date we have observed 25 of the 346 clusters found by Bica and Dutra (2003). We constructed near-IR color-color diagrams to identify PMS stars in the clusters and examined the distribution of these stars in the cluster field. In this proceeding we show the results concerning three candidates of our survey.

## Introduction

In the last decades infrared imaging techniques have evolved rapidly. Large format IR cameras now enable us to capture optically obscured clusters in nebular or molecular clouds, the place of star birth. These star clusters are often referred to as "embedded clusters" (Lada & Lada 1991). Since these young clusters contain almost all of their initial members, they are ideal targets of stellar evolution studies. PMS stars often show large NIR excess due to circumstellar disks (Kenyon et al., 1996). The excess moves these stars to the right side of the reddening band in the $(J-H)$–$(H-K)$ diagram. We have studied 25 of the 346 newly discovered embedded clusters by Bica & Dutra (2003), who did a systematic search of the 2MASS database to find new embedded clusters, close to optical and radio nebulae. The preliminary results of the analysis are presented in this proceeding.

## Observations and data reduction

The JHK observations were carried out on two nights in November, 2003. The infrared images were taken with the STELIRCAM IR camera mounted on the $1.2$ m telescope at Fred Lawrence Whipple Observatory at Mt. Hopkins.

The fields of the clusters were covered with a $3 \times 3$ mosaic, which gave a $7' \times 7'$ field of view. Image reduction was done with the STELIRCAM pipeline of Bill Wyatt which uses the procedures described in Barsony et al. (1997). The combination of the individual frames was performed with the `.drizzle` task in the *STSDAS* package. We used SExtractor (Bertin & Arnouts, 1996) to select sources in the survey and the *DAOPHOT* package to perform aperture photometry on the identified sources.

We transformed our data into the standard system using the IR standars of Elias et al. (1982). We detected a small ($0\overset{m}{.}05$) systematic zero point shift between 2MASS and our data in some of the images. The origin of this difference is yet unknown, therefore, we have shifted our data to the 2MASS system until the source of this inconsistency is cleared.

## Results

### BFS31

BFS31 was first catalogued by Blitz, Fich & Stark (1982). Since then no observation has been made of this HII region. Chan & Fich (1995) estimated the distance of the cluster to $3.26 \pm 0.91$ kpc.

Most of the stars are within the ZAMS reddening band of the color-color diagram (Figure 1). The slope of the reddening law of $1.70$ was adopted from Rieke & Lebofsky (1985) for the CIT system. The concentration of the NIR sources confirms the existence of the embedded cluster (Figure 1). On the CMD (Figure 2a) we have plotted the ZAMS of Siess et al. (2000) shifted with the above distance modulus and the estimated color excess of $E(H-K) = 0\overset{m}{.}22$.

Close to the center allmost all stars are located at the PMS position of the CMD. This might indicate an evapourating outer layer of MS stars and an inner core of star birth at BFS31.

### Sh2-209

To date only a few observations were made of Sh2-209, and none in the near-infrared. Ghosh et al. (2001) observed the evolved HII region in the far infrared. The nebulosity is also visible in optical wavebands. The associated molecular cloud extends to a distance of $\sim 14'$ (Blitz, Fich & Stark, 1982).

The detection of an $H_2O$ maser source by Cesaroni et al. (1988) confirms the ongoing star formation. Considering its large distance (12 kpc, Ghosh et al., 2001) and angular diameter, this complex has an actual size of $\sim 20$ pc.

We detected that more than 60 % of the stars in the field has NIR excess on the color-color diagram (Figure 2c). These stars concentrate in the central part of the coordinate plot (Figure 2d), which confirms the existence of an embedded cluster in this HII cloud.

Shifting the ZAMS to the distance of 12 kpc, the top of the ZAMS just reaches the bottom of our CMD (Figure 2b). This may indicate that the actual distance is less than 12 kpc. Note that the color excess is also uncertain which renders the distance estimate quite difficult.

## W3

W3 is a well studied HII cloud with known star forming activity (Ojha et al., 2004). Megeath et al. (1996) found that the main region of the cluster has an age range of 0.3–1 Myr. Bica & Dutra (2003) identified three different HII regions in the area.

We have identified 60 stars with NIR excess, which can be seen on the color-color diagram (Figure 3c). The density of these stars is larger at the center of the nebula (Figure 3d). A lot of main-sequence stars are visible at the southern part of the cluster, which may indicate an age difference between the northern and southern parts. This would support the result of Megeath et al. (1996).

In figure 3a and 3b we plotted the ZAMS of Siess et al. (2000) with the the distance modulus of $d = 1.83 \pm 0.14$ kpc (Imai et al., 2000) assuming $E(H-K) = 0\overset{m}{.}15$. Figure 3a & 3b shows the seperate CMDs of the three regions of Bica & Dutra (region1 - square, region 2 - cross, region 3 - triangle). Region 1 can be well seperated on the CMD, but the stars from region 2 and 3 are mixed. However, we can distinguish more then 3 regions. The seperation in the CMD's can imply many explanations. The regions can be seperated in space, and in age. On the other hand we might see the effect of patchy extinction. Ojha et al (2004) found that the individual extinction ranges from $0^{m}$ to $24^{m}$, which would give an uncertainty in the luminosty estimates. However the conclusion of age difference of Megeath et al. (1996) gives the the most reasonable anwser.

## Acknowledgments

This work has been supported by the OM FKFP Grant 0010/2001 and the Hungarian OTKA Grants T042509 and T34584. The NASA ADS Abstract Service was used to access data and references. The research has made use of the SIMBAD database, operated at CDS-Strasbourg, France.

## References


Barsony, M., Kenyon, S., Lada, E. A., Teuben, P. J.: 1997, *Astrophysical Journal Supplement*, **112**, 109.

Bertin, E., Arnouts, S.: 1996, *Astronomy & Astrophysics Supplement Series*, **117**, 393.

Bessel, M. S., Brett, J. M.: 1988, *Astronomical Society of the Pacific, Publications*, **100**, 1134-1151.

Bica, E., Dutra, C. M., Soares, J., Barbuy, B.: 2003, *Astronomy & Astrophysics*, **404**, 223-232.

Blitz, L.,Fich, M., Stark, A. A.,: 1982, *Astrophysical Journal Supplement Series*, **49**, 183-206.

Cesaroni et al.: 1988, *Astronomy & Astrophysics Supplement Series*, **76**, 445-458.

Chan, G., Fich, M.: 1995, *Astronomical Journal*, **109**, 2611.

Elias, J. H., Frogel, J. A., Matthews, K., Neugebauer, G.: 1982, *Astronomical Journal*, **87**, 1029-1034.

Ghosh, S. K. et al.: 2001, *Journal of Astrophysics and Astronomy*, **22**, 173-185.

Imai, H. et al.: 2000, *Astrophysical Journal*, **538**, 751-765.

Kenyon, S. J., Yi, I., & Hartmann, L.: 1996, *Astrophysical Journal*, **462**, 439.

Lada, Charles J. & Lada Elizabeth A.: 1991, *The formation and evolution of star clusters*

Megeath, S. T. et al.: 1996, *Astronomy & Astrophysics*, **307**, 775-790.

Meyer, R. M. & Calvet, N.: 1997, *The Astronomical Journal*, **11**, 1.

Ojha, D. K. et al.: 2004, *The Astrophysical Journal*, **608**, 2, 797-808.

Rieke, G. H. & Lebofsky M. J.:1985, *The Astrophysical Journal*, **288**, 618-621.

Siess, L., Dufour, E., Forestini, M.: 2000, *Astronomy & Astrophysics*, **358**, 593.


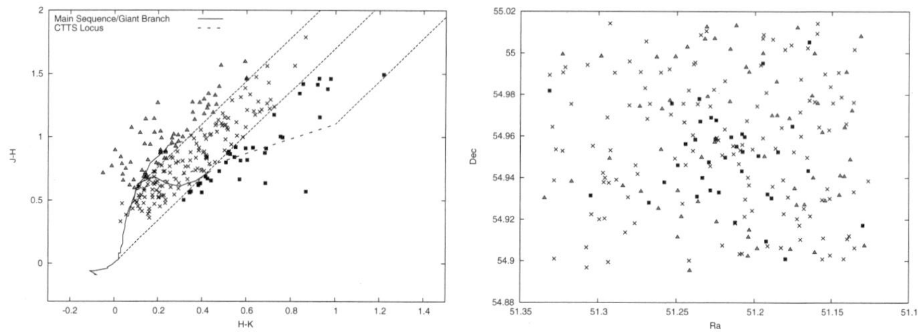


*Figure 1.* Panel (a) shows the color-color diagram of BFS31. Different colors represent stars of different bands (square - PMS stars, cross - MS stars, triangle - stars above the reddening band). The position of the CTTS Locus (dashed line ) was adopted from Meyer (1997). The colors of giant stars were taken from Bessel & Brett (1988) (solid line). The slope of the reddening bands was adopted from Rieke & Lebofsky (1985). Panel (b) shows the distribution of these stars in the field of view (the coding is the same as in panel (a)).

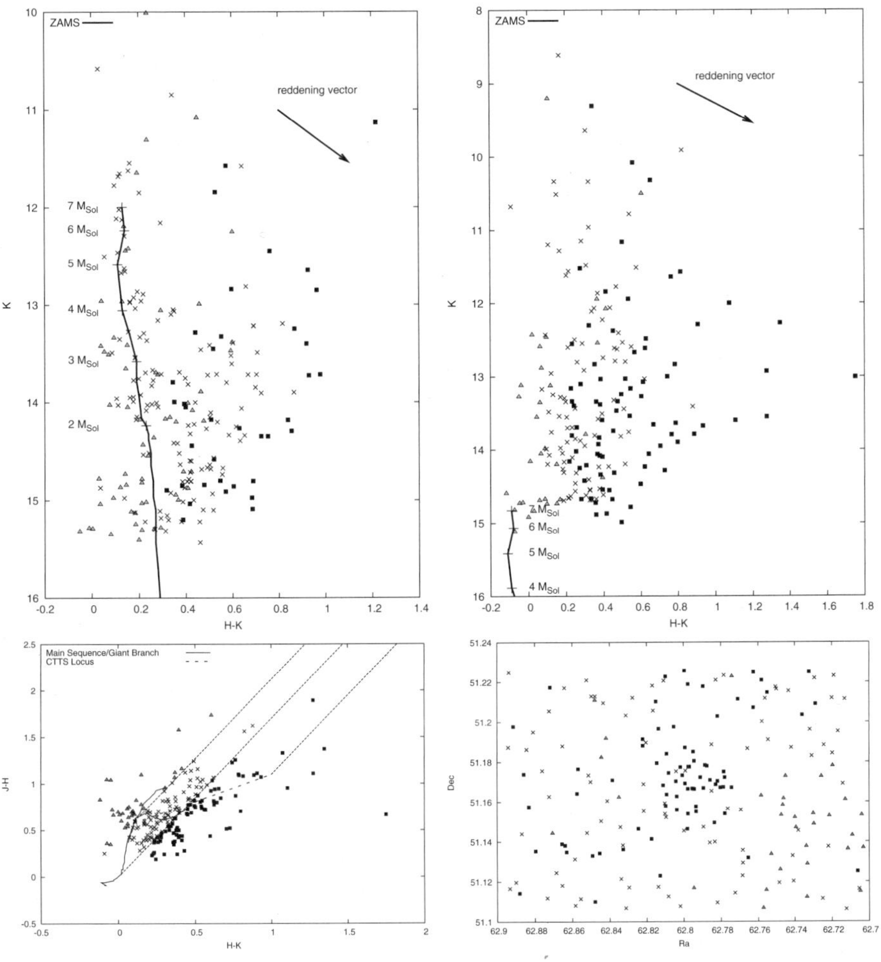


*Figure 2.* Panel (a) and (b) shows color-magnitude diagrams of BFS31 and Sh2-209. Different symbols represent stars of different bands according to the color-color diagrams (Figure 1 and panel (c)). The symbol codings are the same as in Figure 1. Panel (c) and (d) shows color-color- and coordinate diagrams of Sh2-209. The symbol and line codings are the same as in Figure 1.

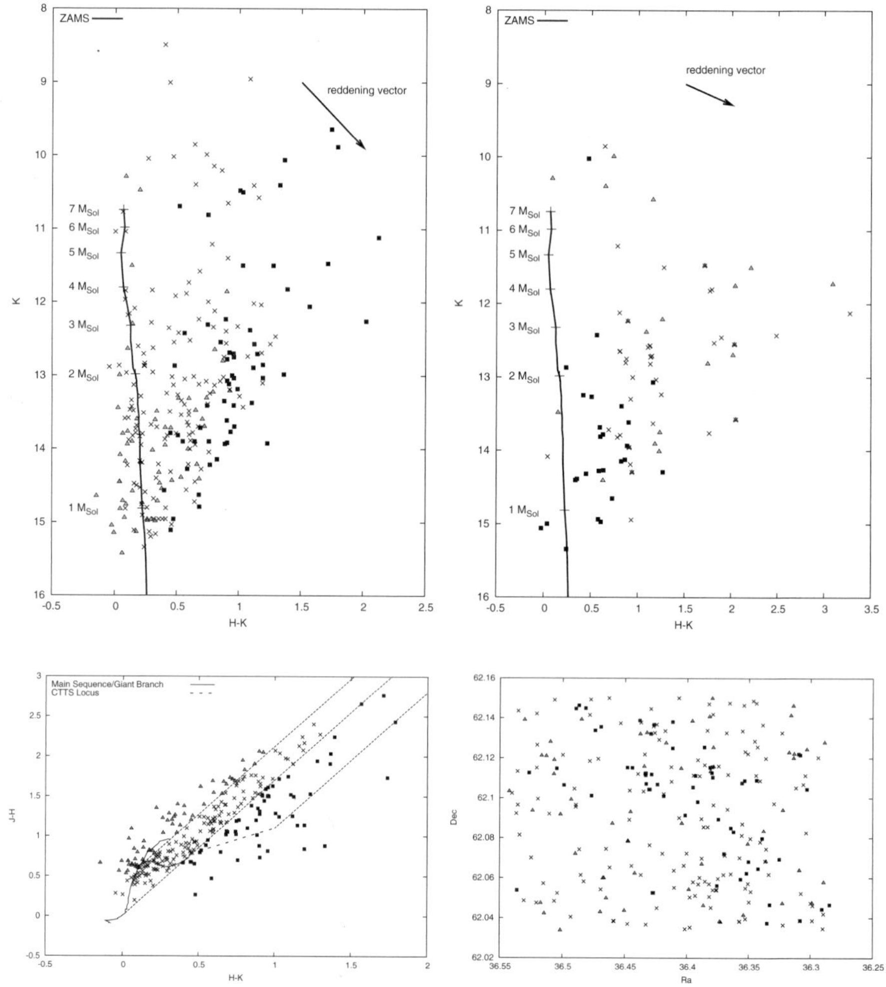


*Figure 3.* Panel (a) & (b) are color-magnitude diagrams of W3. Different symbols in panel (a) represent stars of different bands according to the color-color diagram (panel b). Panel (b) shows the separate CMDs of the three regions defined by Bica & Dutra (square - BD $54^c$, cross - BD $55^c$, triangle - BD $56^c$). Panel (d) shows the distribution of the stars in the field of view. The symbol and line codings are the same as in Figure 1.

# Index

# Astrophysics and Space Science Library

Volume 330: ***The Multinational History of Strasbourg Astronomical Observatory,*** edited by A. Heck. Hardbound ISBN 1-4020-3643-4, June 2005

Volume 329: ***Starbursts – From 30 Doradus to Lyman Break Galaxies,*** edited by R. de Grijs, R.M. González Delgado. Hardbound ISBN 1-4020-3538-1, May 2005

Volume 327: ***The Initial Mass Function 50 Years Later***, edited by E. Corbelli, F. palla, H. Zinnecker. Hardbound ISBN 1-4020-3406-7, June 2005

Volume 325: ***Kristian Birkeland – The First Space Scientist,*** by A. Egeland, W.J. Burke. Hardbound ISBN 1-4020-3293-5, April 2005

Volume 324: ***Cores to Clusters – Star Formation with next Generation Telescopes,*** edited by M.S. Nanda Kumar, M. Tafalla, P. Caselli. Hardbound ISBN 0-387-26322-5, October 2005

Volume 323: ***Recollections of Tucson Operations,*** by M.A. Gordon. Hardbound ISBN 1-4020-3235-8, December 2004

Volume 322: ***Light Pollution Handbook,*** by K. Narisada, D. Schreuder Hardbound ISBN 1-4020-2665-X, November 2004

Volume 321: ***Nonequilibrium Phenomena in Plasmas***, edited by A.S. Shrama, P.K. Kaw. Hardbound ISBN 1-4020-3108-4, December 2004

Volume 320: ***Solar Magnetic Phenomena,*** edited by A. Hanslmeier, A. Veronig, M. Messerotti. Hardbound ISBN 1-4020-2961-6, December 2004

Volume 319: ***Penetrating Bars through Masks of Cosmic Dust***, edited by D.L. Block, I. Puerari, K.C. Freeman, R. Groess, E.K. Block. Hardbound ISBN 1-4020-2861-X, December 2004

Volume 318: ***Transfer of Polarized light in Planetary Atmospheres***, by J.W. Hovenier, J.W. Domke, C. van der Mee. Hardbound ISBN 1-4020-2855-5. Softcover ISBN 1-4020-2889-X, November 2004

Volume 317: ***The Sun and the Heliosphere as an Integrated System,*** edited by G. Poletto, S.T. Suess. Hardbound ISBN 1-4020-2830-X, November 2004

Volume 316: ***Civic Astronomy - Albany's Dudley Observatory, 1852-2002,*** by G. Wise Hardbound ISBN 1-4020-2677-3, October 2004

Volume 315: ***How does the Galaxy Work - A Galactic Tertulia with Don Cox and Ron Reynolds,*** edited by E. J. Alfaro, E. Pérez, J. Franco
Hardbound ISBN 1-4020-2619-6, September 2004

Volume 314: ***Solar and Space Weather Radiophysics- Current Status and Future Developments,*** edited by D.E. Gary and C.U. Keller
Hardbound ISBN 1-4020-2813-X, August 2004

Volume 313: ***Adventures in Order and Chaos***, by G. Contopoulos. Hardbound ISBN 1-4020-3039-8, January 2005

Volume 312: ***High-Velocity Clouds,*** edited by H. van Woerden, U. Schwarz, B. Wakker
Hardbound ISBN 1-4020-2813-X, September 2004

Volume 311: ***The New ROSETTA Targets- Observations, Simulations and Instrument Performances***, edited by L. Colangeli, E. Mazzotta Epifani, P. Palumbo
Hardbound ISBN 1-4020-2572-6, September 2004

Volume 310: ***Organizations and Strategies in Astronomy 5,*** edited by A. Heck
Hardbound ISBN 1-4020-2570-X, September 2004

Volume 309: ***Soft X-ray Emission from Clusters of Galaxies and Related Phenomena***, edited by R. Lieu and J. Mittaz
Hardbound ISBN 1-4020-2563-7, September 2004

Volume 308: ***Supermassive Black Holes in the Distant Universe,*** edited by A.J. Barger
Hardbound ISBN 1-4020-2470-3, August 2004

Volume 307: ***Polarization in Spectral Lines***, by E. Landi Degl'Innocenti and M. Landolfi
Hardbound ISBN 1-4020-2414-2, August 2004

Volume 306: ***Polytropes – Applications in Astrophysics and Related Fields,*** by G.P. Horedt
Hardbound ISBN 1-4020-2350-2, September 2004

Volume 305: ***Astrobiology: Future Perspectives,*** edited by P. Ehrenfreund, W.M. Irvine, T. Owen, L. Becker, J. Blank, J.R. Brucato, L. Colangeli, S. Derenne, A. Dutrey, D. Despois, A. Lazcano, F. Robert
Hardbound ISBN 1-4020-2304-9, July 2004
Paperback ISBN 1-4020-2587-4, July 2004

Volume 304: ***Cosmic Gammy-ray Sources,*** edited by K.S. Cheng and G.E. Romero
Hardbound ISBN 1-4020-2255-7, September 2004

Volume 303: ***Cosmic rays in the Earth's Atmosphere and Underground***, by L.I, Dorman
Hardbound ISBN 1-4020-2071-6, August 2004

Volume 302:***Stellar Collapse,*** edited by Chris L. Fryer
Hardbound, ISBN 1-4020-1992-0, April 2004

Volume 301: ***Multiwavelength Cosmology***, edited by Manolis Plionis
Hardbound, ISBN 1-4020-1971-8, March 2004

Volume 300:***Scientific Detectors for Astronomy,*** edited by Paola Amico, James W. Beletic, Jenna E. Beletic
Hardbound, ISBN 1-4020-1788-X, February 2004

Volume 299: ***Open Issues in Local Star Fomation,*** edited by Jacques Lépine, Jane Gregorio-Hetem
Hardbound, ISBN 1-4020-1755-3, December 2003

Volume 298: ***Stellar Astrophysics - A Tribute to Helmut A. Abt,*** edited by K.S. Cheng, Kam Ching Leung, T.P. Li
Hardbound, ISBN 1-4020-1683-2, November 2003

Volume 297: ***Radiation Hazard in Space,*** by Leonty I. Miroshnichenko
Hardbound, ISBN 1-4020-1538-0, September 2003

Volume 296: ***Organizations and Strategies in Astronomy, volume 4,*** edited by André Heck
Hardbound, ISBN 1-4020-1526-7, October 2003

Volume 295: ***Integrable Problems of Celestial Mechanics in Spaces of Constant Curvature,*** by T.G. Vozmischeva
Hardbound, ISBN 1-4020-1521-6, October 2003

Volume 294: ***An Introduction to Plasma Astrophysics and Magnetohydrodynamics,*** by Marcel Goossens
Hardbound, ISBN 1-4020-1429-5, August 2003
Paperback, ISBN 1-4020-1433-3, August 2003

Volume 293: ***Physics of the Solar System,*** by Bruno Bertotti, Paolo Farinella, David Vokrouhlický
Hardbound, ISBN 1-4020-1428-7, August 2003
Paperback, ISBN 1-4020-1509-7, August 2003

Volume 292: ***Whatever Shines Should Be Observed,*** by Susan M.P. McKenna-Lawlor
Hardbound, ISBN 1-4020-1424-4, September 2003

Volume 291: ***Dynamical Systems and Cosmology***, by Alan Coley
Hardbound, ISBN 1-4020-1403-1, November 2003

Volume 290: ***Astronomy Communication,*** edited by André Heck, Claus Madsen
Hardbound, ISBN 1-4020-1345-0, July 2003

Volume 287/8/9: ***The Future of Small Telescopes in the New Millennium,*** edited by Terry D. Oswalt
Hardbound Set only of 3 volumes, ISBN 1-4020-0951-8, July 2003

Volume 286: ***Searching the Heavens and the Earth: The History of Jesuit Observatories,*** by Agustín Udías
Hardbound, ISBN 1-4020-1189-X, October 2003

Volume 285: ***Information Handling in Astronomy - Historical Vistas,*** edited by André Heck
Hardbound, ISBN 1-4020-1178-4, March 2003

Volume 284: ***Light Pollution: The Global View,*** edited by Hugo E. Schwarz
Hardbound, ISBN 1-4020-1174-1, April 2003

Volume 283: ***Mass-Losing Pulsating Stars and Their Circumstellar Matter,*** edited by Y. Nakada, M. Honma, M. Seki
Hardbound, ISBN 1-4020-1162-8, March 2003

Volume 282: ***Radio Recombination Lines,*** by M.A. Gordon, R.L. Sorochenko
Hardbound, ISBN 1-4020-1016-8, November 2002

Volume 281: ***The IGM/Galaxy Connection,*** edited by Jessica L. Rosenberg, Mary E. Putman
Hardbound, ISBN 1-4020-1289-6, April 2003

Volume 280: ***Organizations and Strategies in Astronomy III,*** edited by André Heck
Hardbound, ISBN 1-4020-0812-0, September 2002

Volume 279: ***Plasma Astrophysics , Second Edition,*** by Arnold O. Benz
Hardbound, ISBN 1-4020-0695-0, July 2002

Volume 278: ***Exploring the Secrets of the Aurora,*** by Syun-Ichi Akasofu
Hardbound, ISBN 1-4020-0685-3, August 2002

Volume 277: ***The Sun and Space Weather,*** by Arnold Hanslmeier
Hardbound, ISBN 1-4020-0684-5, July 2002

Volume 276: ***Modern Theoretical and Observational Cosmology***, edited by Manolis Plionis, Spiros Cotsakis
Hardbound, ISBN 1-4020-0808-2, September 2002

Volume 275: ***History of Oriental Astronomy***, edited by S.M. Razaullah Ansari
Hardbound, ISBN 1-4020-0657-8, December 2002

Volume 274: ***New Quests in Stellar Astrophysics: The Link Between Stars and Cosmology***, edited by Miguel Chávez, Alessandro Bressan, Alberto Buzzoni,Divakara Mayya
Hardbound, ISBN 1-4020-0644-6, June 2002

Volume 273: ***Lunar Gravimetry***, by Rune Floberghagen
Hardbound, ISBN 1-4020-0544-X, May 2002

Volume 272:***Merging Processes in Galaxy Clusters***, edited by L. Feretti, I.M. Gioia, G. Giovannini
Hardbound, ISBN 1-4020-0531-8, May 2002

Volume 271: ***Astronomy-inspired Atomic and Molecular Physics***, by A.R.P. Rau
Hardbound, ISBN 1-4020-0467-2, March 2002

Volume 270: ***Dayside and Polar Cap Aurora***, by Per Even Sandholt, Herbert C. Carlson, Alv Egeland
Hardbound, ISBN 1-4020-0447-8, July 2002

Volume 269: ***Mechanics of Turbulence of Multicomponent Gases***, by Mikhail Ya. Marov, Aleksander V. Kolesnichenko
Hardbound, ISBN 1-4020-0103-7, December 2001

Volume 268: ***Multielement System Design in Astronomy and Radio Science***, by Lazarus E. Kopilovich, Leonid G. Sodin
Hardbound, ISBN 1-4020-0069-3, November 2001

Volume 267: ***The Nature of Unidentified Galactic High-Energy Gamma-Ray Sources***, edited by Alberto Carramiñana, Olaf Reimer, David J. Thompson
Hardbound, ISBN 1-4020-0010-3, October 2001

Volume 266: ***Organizations and Strategies in Astronomy II***, edited by André Heck
Hardbound, ISBN 0-7923-7172-0, October 2001

Volume 265: ***Post-AGB Objects as a Phase of Stellar Evolution***, edited by R. Szczerba, S.K. Górny
Hardbound, ISBN 0-7923-7145-3, July 2001

Volume 264: ***The Influence of Binaries on Stellar Population Studies,*** edited by Dany Vanbeveren
Hardbound, ISBN 0-7923-7104-6, July 2001

Volume 262: ***Whistler Phenomena - Short Impulse Propagation,*** by Csaba Ferencz, Orsolya E. Ferencz, Dániel Hamar, János Lichtenberger
Hardbound, ISBN 0-7923-6995-5, June 2001

Volume 261: ***Collisional Processes in the Solar System,*** edited by Mikhail Ya. Marov, Hans Rickman
Hardbound, ISBN 0-7923-6946-7, May 2001

Volume 260: ***Solar Cosmic Rays,*** by Leonty I. Miroshnichenko
Hardbound, ISBN 0-7923-6928-9, May 2001

For further information about this book series we refer you to the following web site:
www.springeronline.com

To contact the Publishing Editor for new book proposals:
Dr. Harry (J.J.) Blom: harry.blom@springer-sbm.com
Sonja Japenga: sonja.japenga@springer-sbm.com